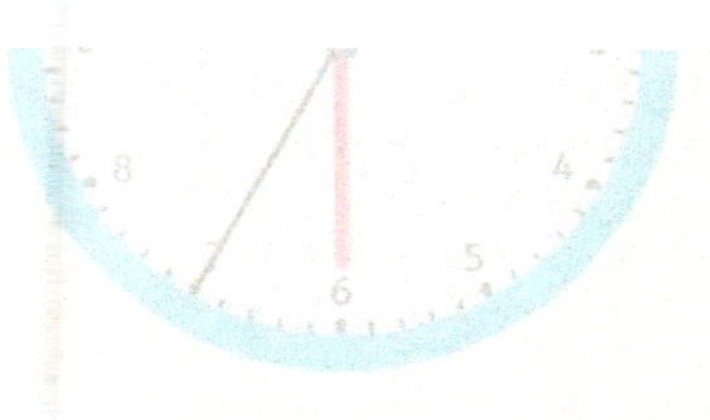

Math

Volume 2

ISBN: 9798419979253

1a

Do you remember?

$\frac{1}{4} + \frac{1}{4} = \frac{?}{?}$

Simplify if possible.

Note taking

1b

Do you remember?

$\frac{3}{7} + \frac{2}{7} = \frac{\square}{\square}$

Simplify if possible.

Note taking

1h

$\frac{1}{4} + \frac{3}{8} = ?$

common denominators

$\frac{\square}{8} + \frac{3}{8} = \frac{\square}{8}$

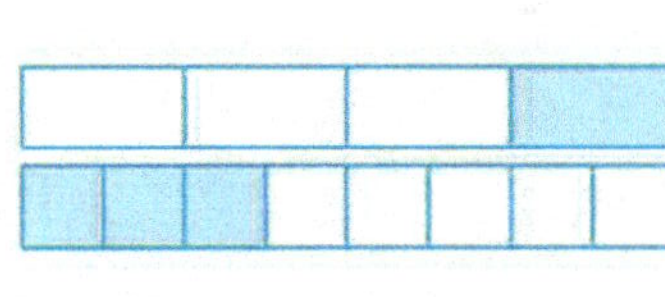

Note taking

1i

Show your work on your paper.

$\frac{7}{12} + \frac{1}{3} = \frac{?}{?}$

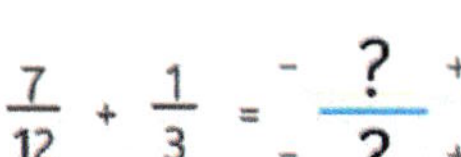

Note taking

1j

Show your work on paper.

Simplify if possible.

$\frac{2}{5} + \frac{1}{10} = \frac{?}{?}$

Note taking

2a

$\frac{1}{2} + \frac{1}{8} = ?$

common denominators

$\frac{\square}{8} + \frac{1}{8} = \frac{\square}{8}$

Note taking

2b

$\frac{3}{8} + \frac{1}{4} = ?$

common denominators

$\frac{3}{8} + \frac{\square}{8} = \frac{\square}{\square}$

Note taking

2c

$\frac{2}{3} + \frac{2}{9} = ?$

common denominators

$\frac{\square}{\square} + \frac{2}{9} = \frac{\square}{9}$

Note taking

2d

$\frac{8}{15} + \frac{1}{3} = ?$

common denominators

$\frac{8}{15} + \frac{\square}{\square} = \frac{\square}{15}$

Note taking

2e

$\frac{2}{5} + \frac{1}{10} = \frac{?}{?}$

Note taking

2f

$\frac{2}{3} + \frac{1}{6} = \frac{?}{?}$

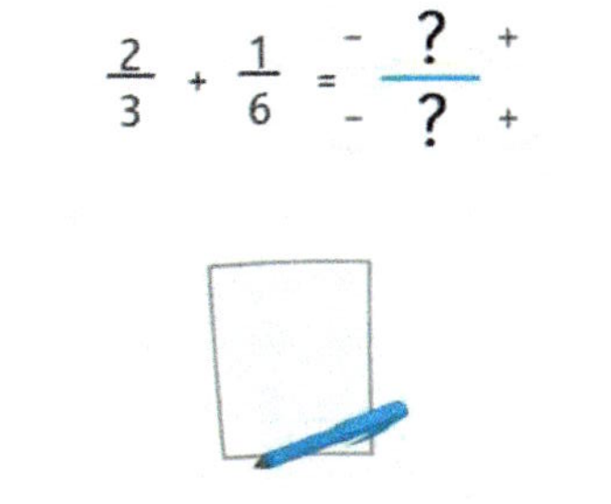

Note taking

2g

Show your work on paper.

Simplify if possible.

$\frac{3}{4} + \frac{3}{16} = \frac{?}{?}$

Note taking

2h

Show your work on paper.

Simplify if possible.

$\frac{7}{12} + \frac{1}{3} = \frac{?}{?}$

Note taking

2i

Show your work on paper.

Simplify if possible.

$\frac{2}{5} + \frac{1}{15} = \frac{?}{?}$

Note taking

2j

Show you work on paper.

Simplify if possible.

$\frac{1}{3} + \frac{5}{12} = \frac{?}{?}$

Note taking

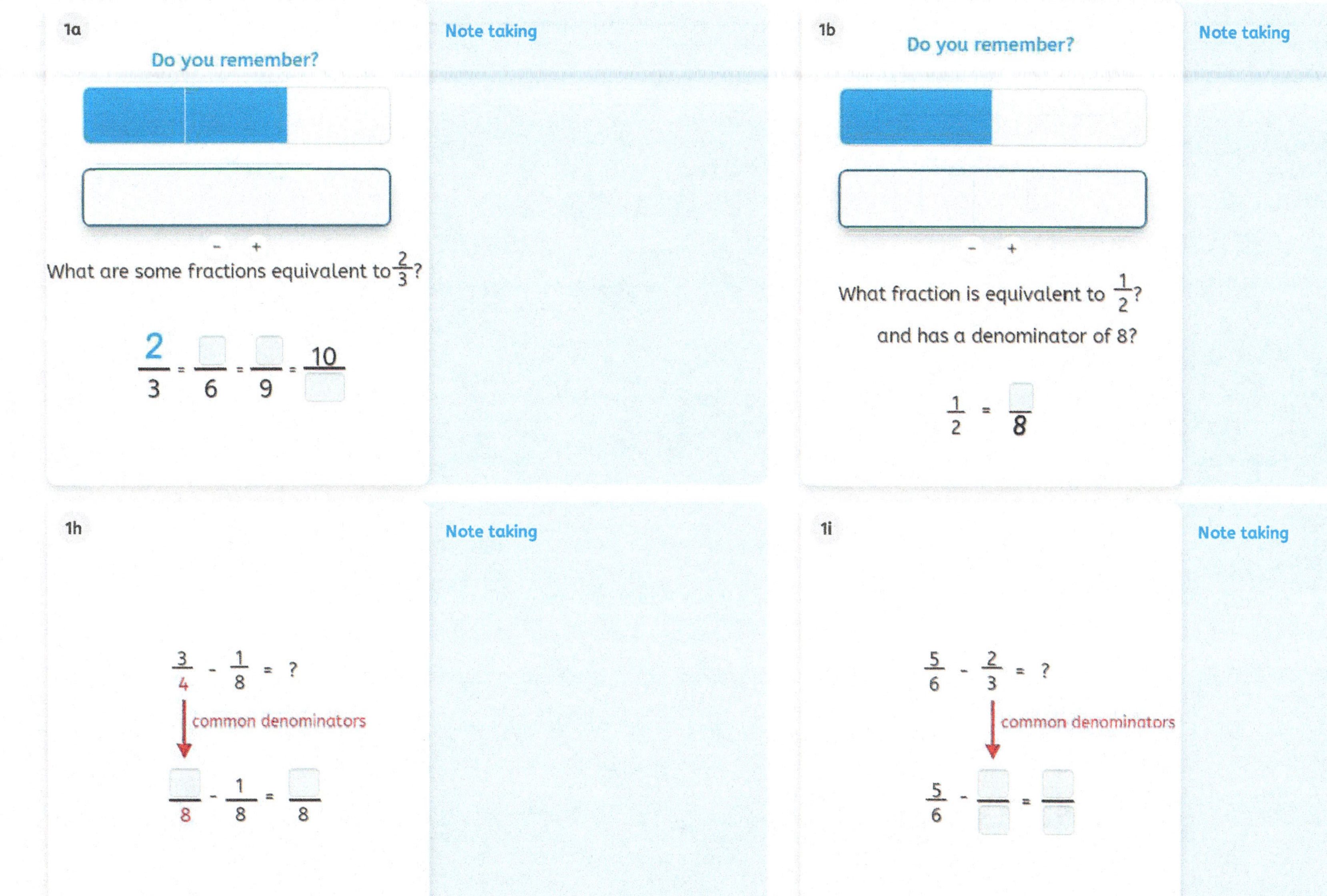
1a
Do you remember?
- +
What are some fractions equivalent to $\frac{2}{3}$?
$\frac{2}{3} = \frac{\square}{6} = \frac{\square}{9} = \frac{10}{\square}$
Note taking
1b
Do you remember?
- +
What fraction is equivalent to $\frac{1}{2}$?
and has a denominator of 8?
$\frac{1}{2} = \frac{\square}{8}$
Note taking
1h
$\frac{3}{4} - \frac{1}{8} = ?$
common denominators
$\frac{\square}{8} - \frac{1}{8} = \frac{\square}{8}$
Note taking
1i
$\frac{5}{6} - \frac{2}{3} = ?$
common denominators
$\frac{5}{6} - \frac{\square}{\square} = \frac{\square}{\square}$
Note taking

1j

Show work on your paper.

Simplify if possible.

$\frac{1}{2} - \frac{1}{8} = \frac{?}{?}$

Note taking

1k

Show work on your paper.

Simplify if possible.

$\frac{11}{12} - \frac{1}{3} = \frac{?}{?}$

Note taking

1l

Show work on your paper.

Simplify if possible.

$\frac{2}{5} - \frac{1}{10} = \frac{?}{?}$

Note taking

2a

$\frac{1}{2} - \frac{1}{8} = ?$

common denominators

$\frac{\square}{8} - \frac{1}{8} = \frac{\square}{8}$

Note taking

2b

$\frac{7}{8} - \frac{3}{4} = ?$

common denominators

$\frac{7}{8} - \frac{\square}{8} = \frac{\square}{\square}$

Note taking

2c

$\frac{2}{3} - \frac{2}{9} = ?$

common denominators

$\frac{\square}{\square} - \frac{2}{9} = \frac{\square}{9}$

Note taking

2d

$\frac{11}{15} - \frac{2}{3} = ?$

common denominators

$\frac{11}{15} - \frac{\square}{\square} = \frac{\square}{\square}$

Note taking

2e

Show work on your paper.

Simplify if possible.

$\frac{2}{5} - \frac{1}{10} =$

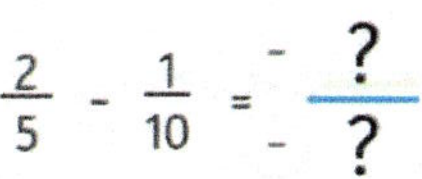

Note taking

2f

Show work on your paper.

Simplify if possible.

$\frac{2}{3} - \frac{1}{6} = \frac{?}{?}$

Note taking

2g

Show work on your paper.

Simplify if possible.

$\frac{3}{4} - \frac{3}{16} = \frac{?}{?}$

Note taking

2h

Show work on your paper.

Simplify if possible.

$\frac{7}{12} - \frac{1}{3} = \frac{?}{?}$

Note taking

2i

Show work on your paper.

Simplify if possible.

$\frac{2}{5} - \frac{1}{15} = \frac{?}{?}$

Note taking

2j

Show work on your paper.

Simplify if possible.

$\frac{5}{12} - \frac{1}{3} = \frac{?}{?}$

Note taking

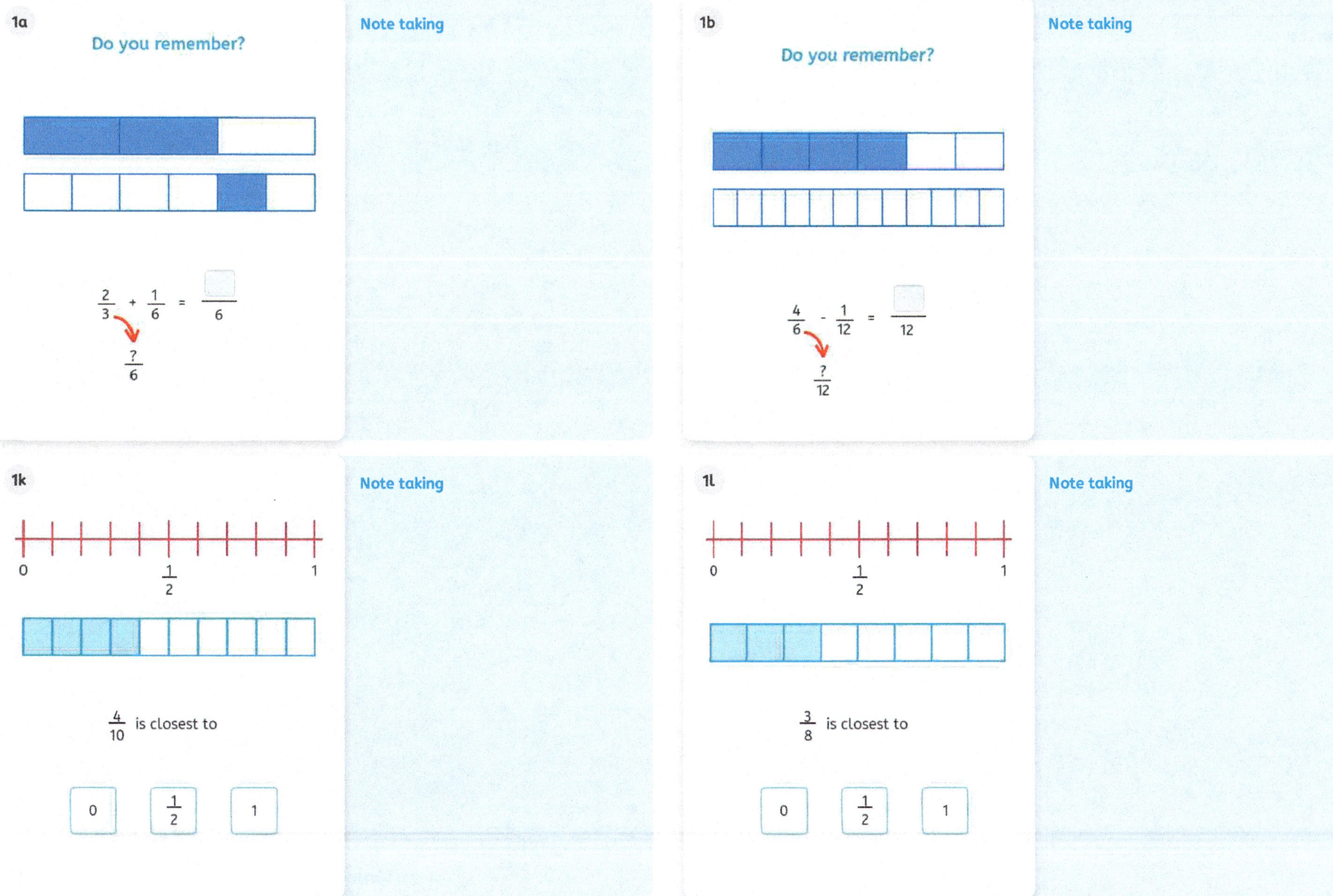
1a
Do you remember?
$\frac{2}{3} + \frac{1}{6} = \frac{\square}{6}$
$\frac{?}{6}$
Note taking
1b
Do you remember?
$\frac{4}{6} - \frac{1}{12} = \frac{\square}{12}$
$\frac{?}{12}$
Note taking
1k
0
$\frac{1}{2}$
1
$\frac{4}{10}$ is closest to
0
$\frac{1}{2}$
1
Note taking
1l
0
$\frac{1}{2}$
1
$\frac{3}{8}$ is closest to
0
$\frac{1}{2}$
1
Note taking

1m

0 $\frac{1}{2}$ 1

$\frac{5}{6}$ is closest to

0 | $\frac{1}{2}$ | 1

Note taking

1n

0 $\frac{1}{2}$ 1

$\frac{3}{8} + \frac{2}{6}$ is closest to

0 | $\frac{1}{2}$ | 1

Note taking

1o

0 $\frac{1}{2}$ 1

$\frac{2}{6} + \frac{1}{10}$ is closest to

0 | $\frac{1}{2}$ | 1

Note taking

1p

Estimate the sum.
Add the rounded numbers.

$\frac{4}{9} + \frac{6}{10} \approx$

Round to $0, \frac{1}{2}$ or 1 → $\frac{1}{2}$

Round to $0, \frac{1}{2}$ or 1 → $\frac{1}{2}$

≈ means approximately

Note taking

1q

0 $\frac{1}{2}$ 1

$\frac{6}{8} - \frac{6}{10}$ is closest to

0 $\frac{1}{2}$ 1

Note taking

1r

0 $\frac{1}{2}$ 1

$\frac{7}{8} - \frac{6}{10}$ is closest to

0 $\frac{1}{2}$ 1

Note taking

1s

Estimate the sum.
Add the rounded numbers.

$\frac{8}{9} + \frac{4}{10} \approx$

Round to $0, \frac{1}{2}$ or 1

Round to $0, \frac{1}{2}$ or 1

≈ means approximately

Note taking

1t

Round the fractions,
and estimate the difference.

$\frac{6}{7} - \frac{1}{5} \approx$

Round to $0, \frac{1}{2}$ or 1

Round to $0, \frac{1}{2}$ or 1

≈ means approximately

Note taking

1u

Round the fractions,
and estimate the difference.

$2\frac{6}{7} - 1\frac{1}{5} \approx \square \frac{\square}{\square}$

≈ means approximately

Note taking

2a

0 $\frac{1}{2}$ 1

$\frac{4}{6}$ is closest to:

0 $\frac{1}{2}$ 1

Note taking

2b

0 $\frac{1}{2}$ 1

$\frac{5}{6}$ is closest to:

0 $\frac{1}{2}$ 1

Note taking

2c

0 $\frac{1}{2}$ 1

$\frac{2}{10} + \frac{2}{6}$ is closest to:

0 $\frac{1}{2}$ 1

Note taking

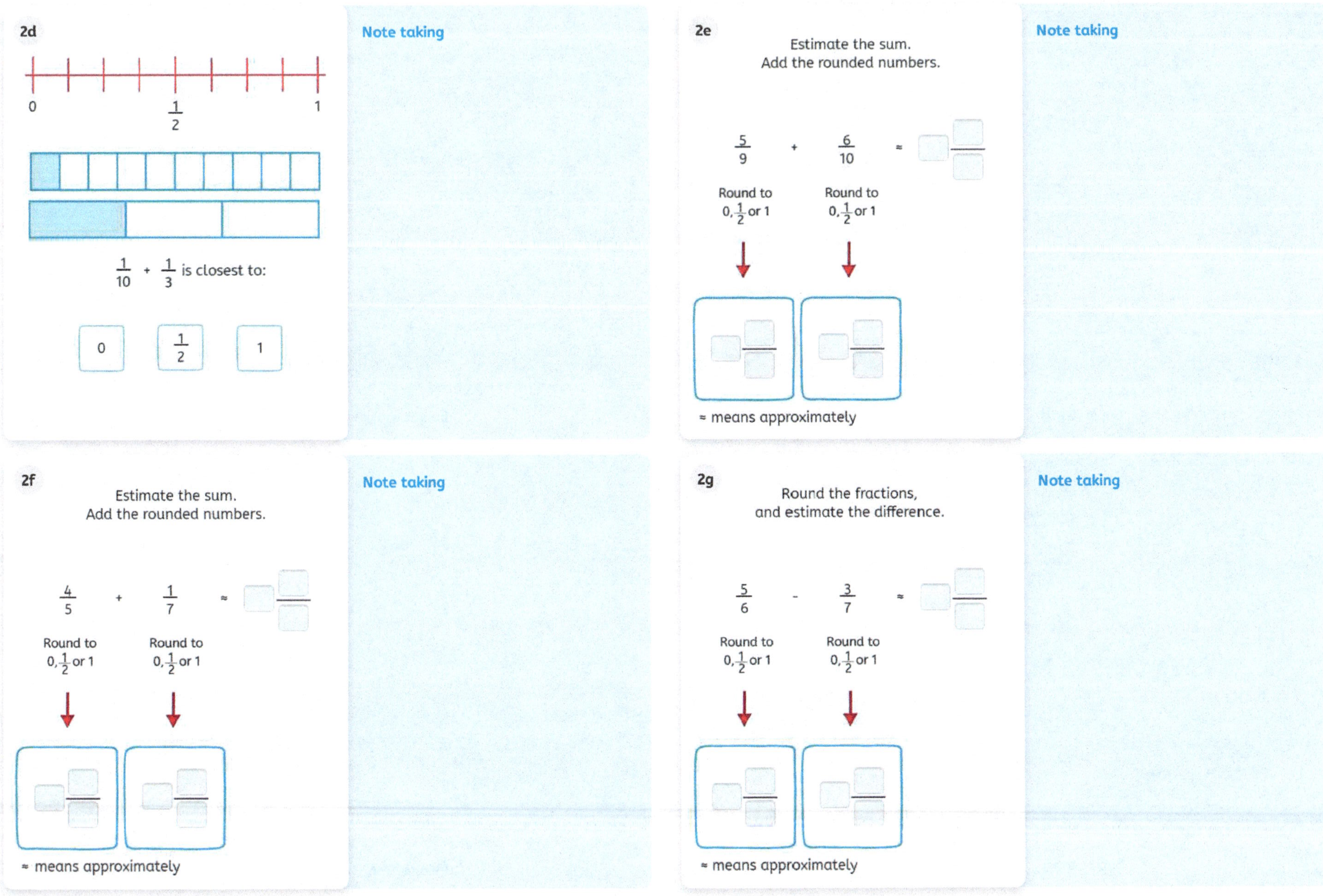

2d

0 $\frac{1}{2}$ 1

$\frac{1}{10} + \frac{1}{3}$ is closest to:

0 $\frac{1}{2}$ 1

Note taking

2e

Estimate the sum.
Add the rounded numbers.

$\frac{5}{9} + \frac{6}{10} \approx$

Round to 0, $\frac{1}{2}$ or 1 — Round to 0, $\frac{1}{2}$ or 1

≈ means approximately

Note taking

2f

Estimate the sum.
Add the rounded numbers.

$\frac{4}{5} + \frac{1}{7} \approx$

Round to 0, $\frac{1}{2}$ or 1 — Round to 0, $\frac{1}{2}$ or 1

≈ means approximately

Note taking

2g

Round the fractions,
and estimate the difference.

$\frac{5}{6} - \frac{3}{7} \approx$

Round to 0, $\frac{1}{2}$ or 1 — Round to 0, $\frac{1}{2}$ or 1

≈ means approximately

Note taking

2h

Round the fractions,
and estimate the difference.

$\frac{6}{10} - \frac{5}{11} \approx$ ☐ $\frac{☐}{☐}$

Round to $0, \frac{1}{2}$ or 1   Round to $0, \frac{1}{2}$ or 1

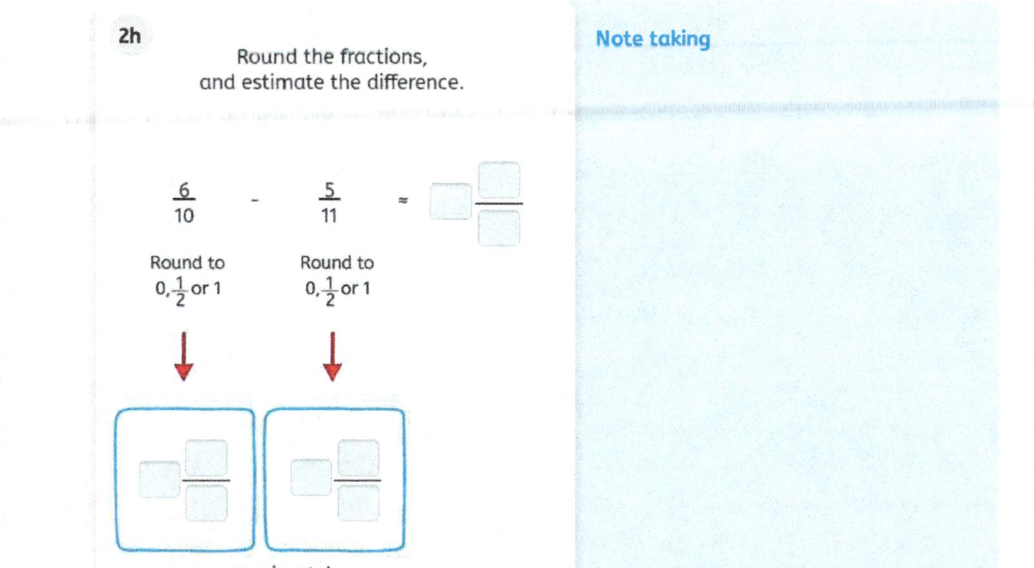

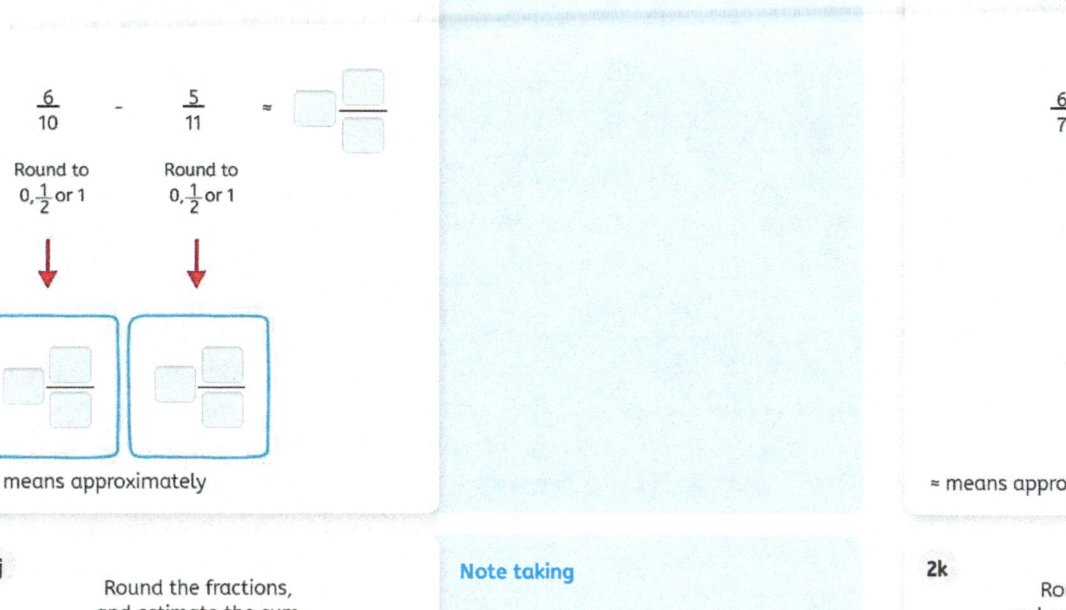

≈ means approximately

Note taking

2i

Round the fractions,
and estimate the sum.

$\frac{6}{7} + \frac{4}{7} \approx$ ☐ $\frac{☐}{☐}$

≈ means approximately

Note taking

2j

Round the fractions,
and estimate the sum.

$\frac{5}{12} + \frac{8}{9} \approx$ ☐ $\frac{☐}{☐}$

≈ means approximately

Note taking

2k

Round the fractions,
and estimate the difference.

$\frac{8}{9} - \frac{1}{6} \approx$ ☐ $\frac{☐}{☐}$

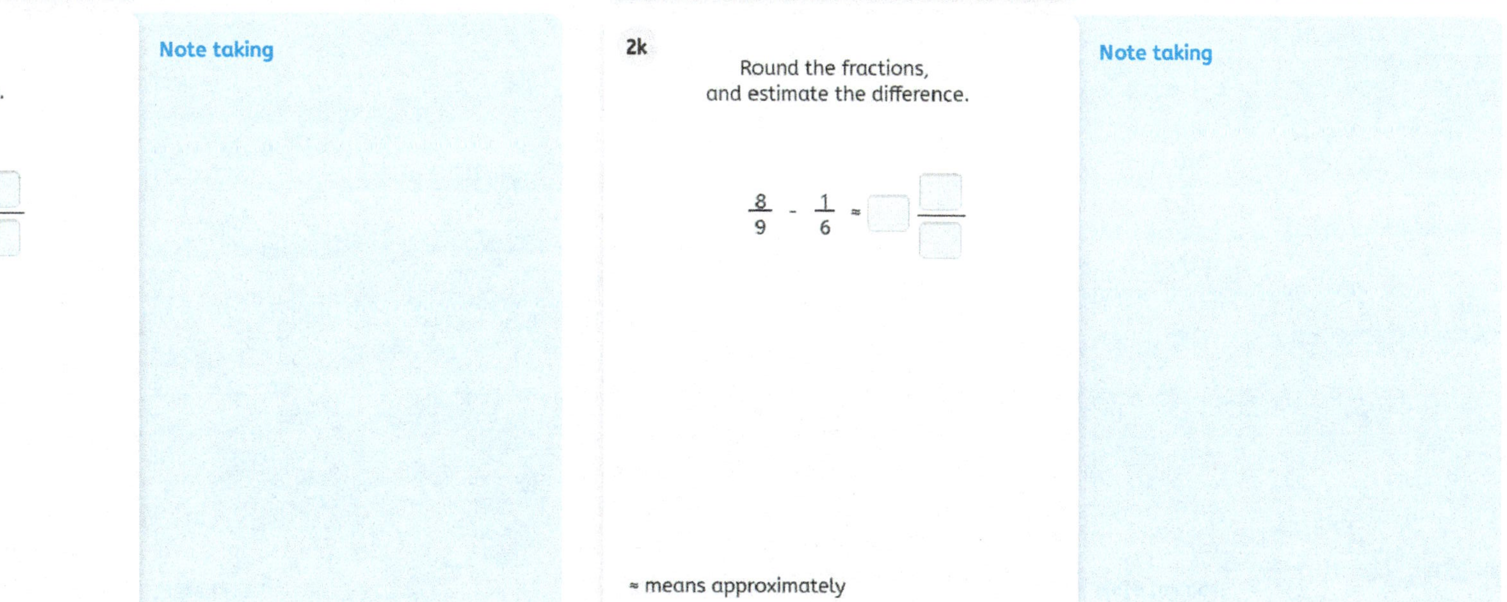

≈ means approximately

Note taking

2l

Round the fractions,
and estimate the difference.

$\frac{11}{12} - \frac{2}{11} \approx \square \frac{\square}{\square}$

≈ means approximately

Note taking

2m

Round the fractions,
and estimate the sum.

$1\frac{5}{6} + 1\frac{5}{9} \approx \square \frac{\square}{\square}$

≈ means approximately

Note taking

2n

Round the fractions,
and estimate the difference.

$4\frac{9}{11} - 2\frac{3}{8} \approx \square \frac{\square}{\square}$

≈ means approximately

Note taking

2o

Paul cut $\frac{3}{5}$ of a yard off a plank.

The plank was $2\frac{1}{8}$ yards long.

About how long is the remaining part?

0 yards	$\frac{1}{2}$ yard	1 yard	$1\frac{1}{2}$ yards
2 yards	$2\frac{1}{2}$ yards	3 yards	$3\frac{1}{2}$ yards

Note taking

2p

Carol brings her lunch in her backpack.

The water weights $1\frac{1}{8}$ pounds, the apple $\frac{2}{5}$ pound and bread with jam $\frac{1}{6}$ pound.

About how much does the lunch weigh?

0 lb	$\frac{1}{2}$ lb	1 lb	$1\frac{1}{2}$ lb
2 lb	$2\frac{1}{2}$ lb	3 lb	$3\frac{1}{2}$ lb

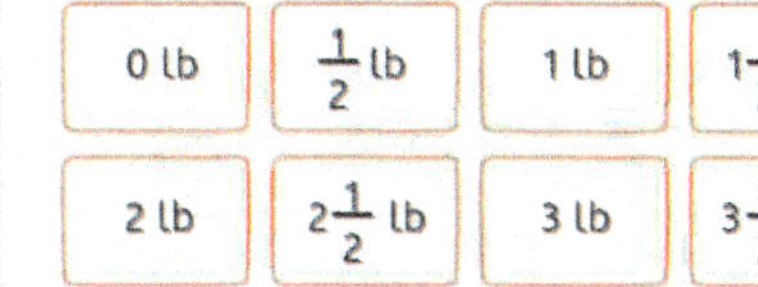

Note taking

1a

Do you remember?

$\frac{3}{4} - \frac{2}{4} = \frac{\square}{4}$

Note taking

1b

Do you remember?

$\frac{1}{4} + \frac{1}{2} = ?$

Tip

$\frac{1}{4} + \frac{\square}{4} = \frac{\square}{4}$

Note taking

1o

$\frac{1}{6} + \frac{1}{5} = ?$

$\frac{\square}{30} + \frac{\square}{30} = \frac{\square}{30}$

Fill in all the boxes.

Note taking

1p

$\frac{5}{7} + \frac{2}{3} = ?$

$\frac{\square}{\square} + \frac{\square}{\square} = \square\frac{\square}{\square}$

Fill in all the boxes.

Note taking

1q

$\frac{1}{2} - \frac{2}{7} = ?$

$\frac{\square}{14} - \frac{\square}{\square} = \frac{\square}{\square}$

Fill in all the boxes.

Note taking

1r

$\frac{1}{6} - \frac{1}{7} = \frac{\square}{\square}$

Note taking

1s

$\frac{5}{8} - \frac{2}{5} = \frac{\square}{\square}$

Note taking

2a

$\frac{1}{4} + \frac{1}{5} = ?$

$\frac{\square}{20} + \frac{\square}{20} = \frac{\square}{20}$

Fill in all the boxes.

Note taking

2b

$\frac{2}{3} - \frac{1}{5} = ?$

$\frac{\square}{15} - \frac{\square}{\square} = \frac{\square}{\square}$

Fill in all the boxes.

Note taking

2c

$\frac{3}{7} - \frac{1}{3} = ?$

$\frac{\square}{\square} - \frac{\square}{\square} = \frac{\square}{\square}$

Fill in all the boxes.

Note taking

2d

$\frac{4}{9} + \frac{1}{2} = ?$

$\frac{\square}{\square} + \frac{\square}{\square} = \frac{\square}{\square}$

Fill in all the boxes.

Note taking

2e

$\frac{2}{3} + \frac{3}{8} = ?$

$\frac{\square}{\square} + \frac{\square}{\square} = \square\frac{\square}{\square}$

Fill in all the boxes.

Note taking

2f

$\frac{1}{4} - \frac{1}{9} = \frac{\square}{\square}$

Note taking

2g

$\frac{5}{6} - \frac{4}{5} = \frac{\square}{\square}$

Note taking

2h

$\frac{2}{7} + \frac{1}{4} = \frac{\square}{\square}$

Note taking

2i

$\frac{4}{9} + \frac{4}{5} = \square\frac{\square}{\square}$

Note taking

2j

$\frac{4}{7} - \frac{3}{8} = \frac{\square}{\square}$

Note taking

1a

Do you remember?

$\frac{1}{6} - \frac{1}{7} = \frac{\square}{\square}$

Note taking

1b

Do you remember?

$\frac{2}{3} + \frac{3}{4} = ?$

$\frac{8}{12} + \frac{9}{12} = \frac{17}{12} = \square\frac{\square}{\square}$

Note taking

1e

drag to fill

$2\frac{3}{8} + 1\frac{1}{2} = \square\frac{\square}{\square}$

Simplify your answer if possible.

Note taking

1f

drag to fill

$2\frac{3}{8} + 1\frac{1}{4} = \square\frac{\square}{\square}$

Simplify your answer if possible.

Note taking

1h Hint

$1\frac{2}{3} + 2\frac{1}{8} = \square\frac{\square}{\square}$

$\square\frac{\square}{24}$ $\square\frac{\square}{24}$

Simplify your answer if possible.

Note taking

1i Hint

$1\frac{3}{4} + 2\frac{3}{8} = \square\frac{\square}{\square}$

$\square\frac{\square}{\square}$ $\square\frac{\square}{\square}$

Simplify your answer if possible.

Note taking

1j

$1\frac{1}{2} + \frac{3}{5} = \square\frac{\square}{\square}$

Note taking

1l

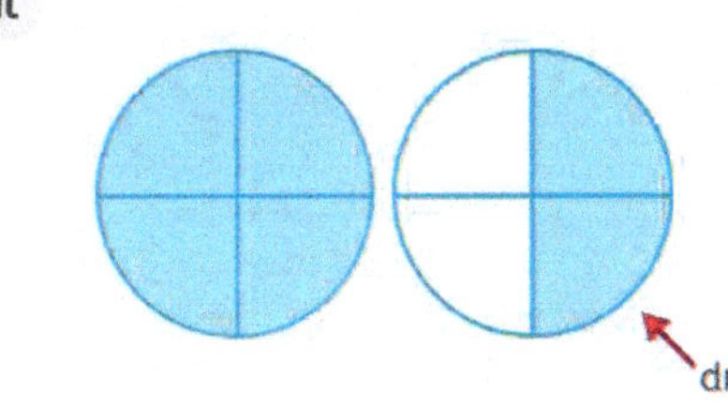

$1\frac{1}{2} - \frac{1}{4} = \square\frac{\square}{\square}$

Simplify your answer if possible.

Note taking

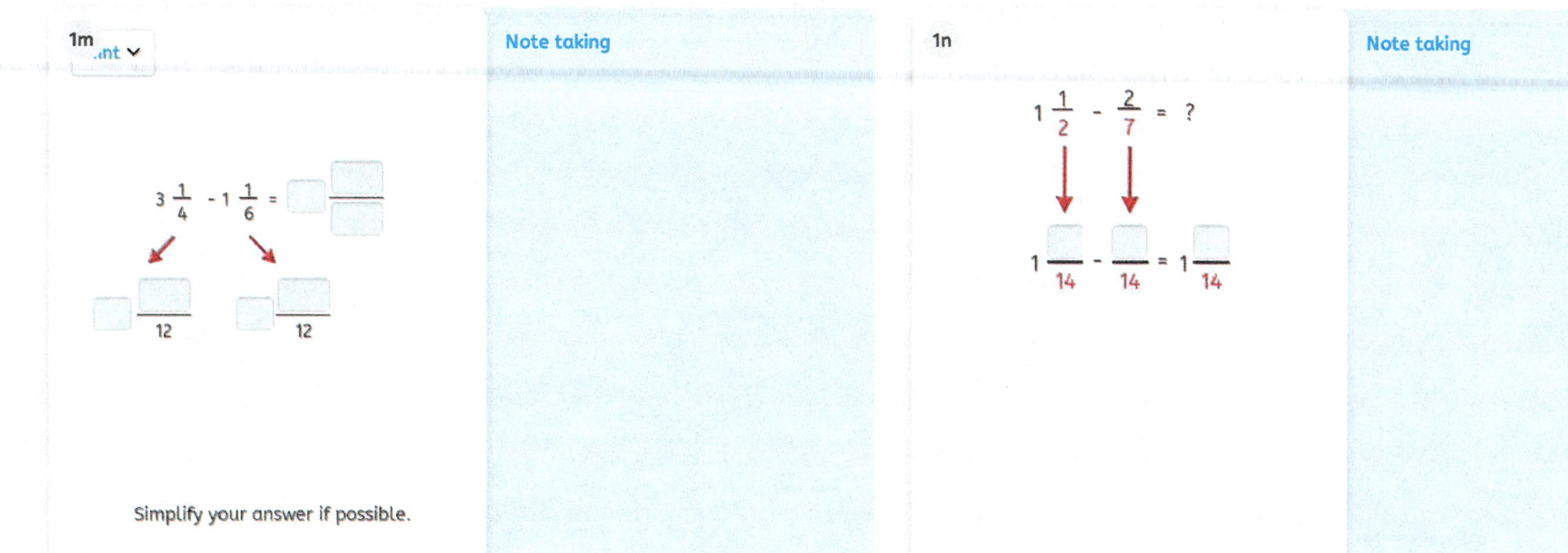

1m Hint

$3\frac{1}{4} - 1\frac{1}{6} =$

$\frac{\square}{12}$ $\frac{\square}{12}$

Simplify your answer if possible.

Note taking

1n

$1\frac{1}{2} - \frac{2}{7} = ?$

$1\frac{\square}{14} - \frac{\square}{14} = 1\frac{\square}{14}$

Note taking

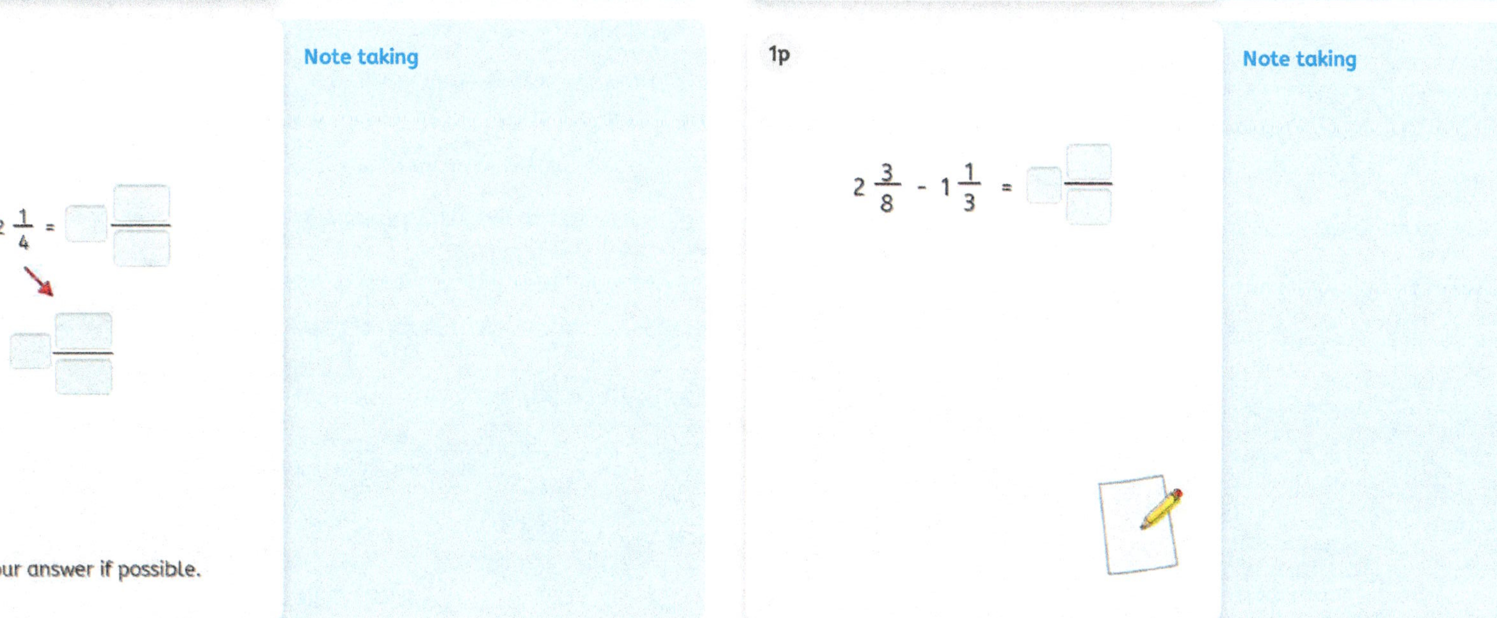

1o Hint

$4\frac{1}{3} - 2\frac{1}{4} =$

Simplify your answer if possible.

Note taking

1p

$2\frac{3}{8} - 1\frac{1}{3} =$

Note taking

2a

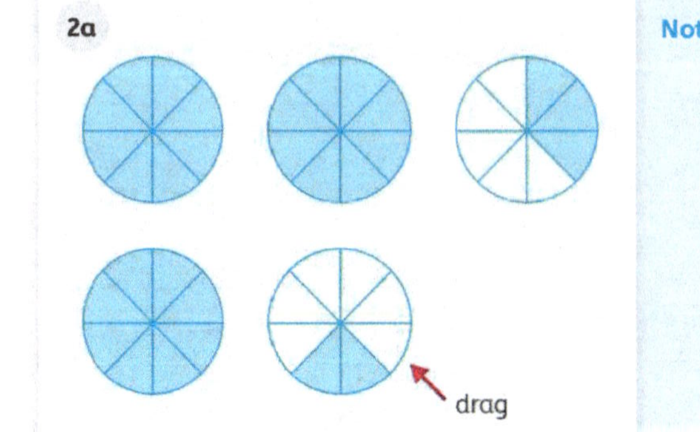

$2\frac{3}{8} + 1\frac{1}{4} = \square\frac{\square}{\square}$

Simplify your answer if possible.

Note taking

2b

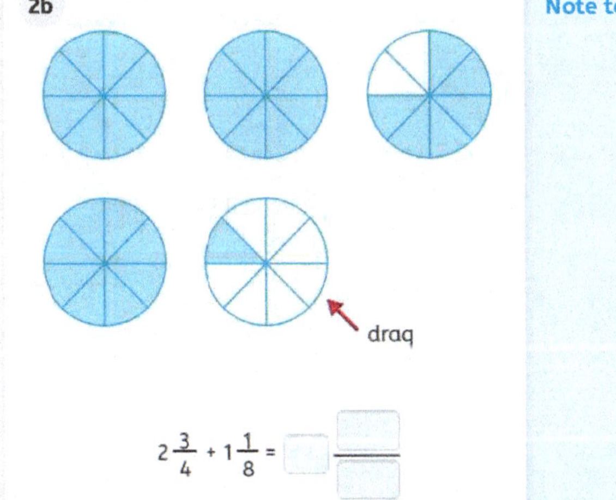

$2\frac{3}{4} + 1\frac{1}{8} = \square\frac{\square}{\square}$

Simplify your answer if possible.

Note taking

2c Hint

$1\frac{2}{3} + 2\frac{1}{6} = \square\frac{\square}{\square}$

$\square\frac{\square}{6}$ $\square\frac{\square}{6}$

Simplify your answer if possible.

Note taking

2d Hint

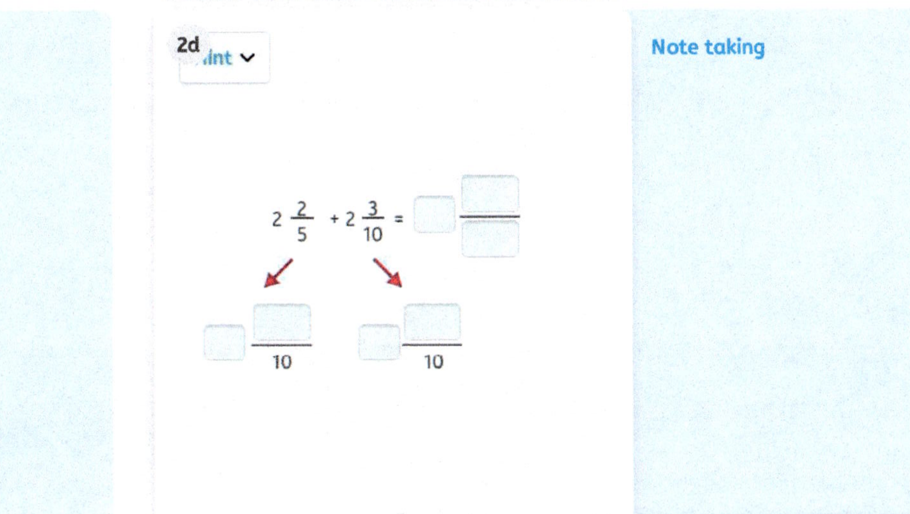

$2\frac{2}{5} + 2\frac{3}{10} = \square\frac{\square}{\square}$

$\square\frac{\square}{10}$ $\square\frac{\square}{10}$

Simplify your answer if possible.

Note taking

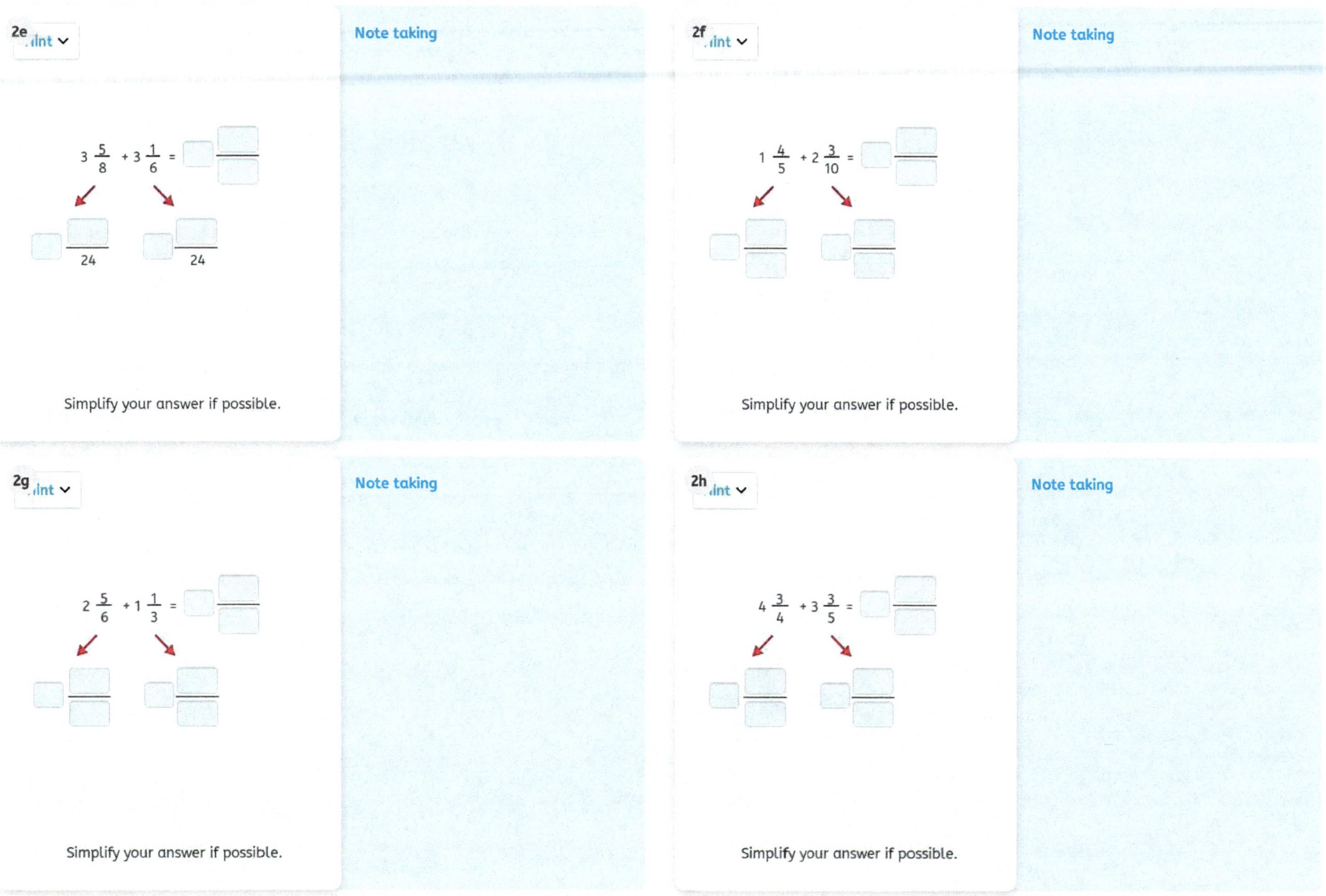

2e Hint

$3\frac{5}{8} + 3\frac{1}{6} = \square\frac{\square}{\square}$

$\square\frac{\square}{24}$ $\square\frac{\square}{24}$

Simplify your answer if possible.

Note taking

2f Hint

$1\frac{4}{5} + 2\frac{3}{10} = \square\frac{\square}{\square}$

$\square\frac{\square}{\square}$ $\square\frac{\square}{\square}$

Simplify your answer if possible.

Note taking

2g Hint

$2\frac{5}{6} + 1\frac{1}{3} = \square\frac{\square}{\square}$

$\square\frac{\square}{\square}$ $\square\frac{\square}{\square}$

Simplify your answer if possible.

Note taking

2h Hint

$4\frac{3}{4} + 3\frac{3}{5} = \square\frac{\square}{\square}$

$\square\frac{\square}{\square}$ $\square\frac{\square}{\square}$

Simplify your answer if possible.

Note taking

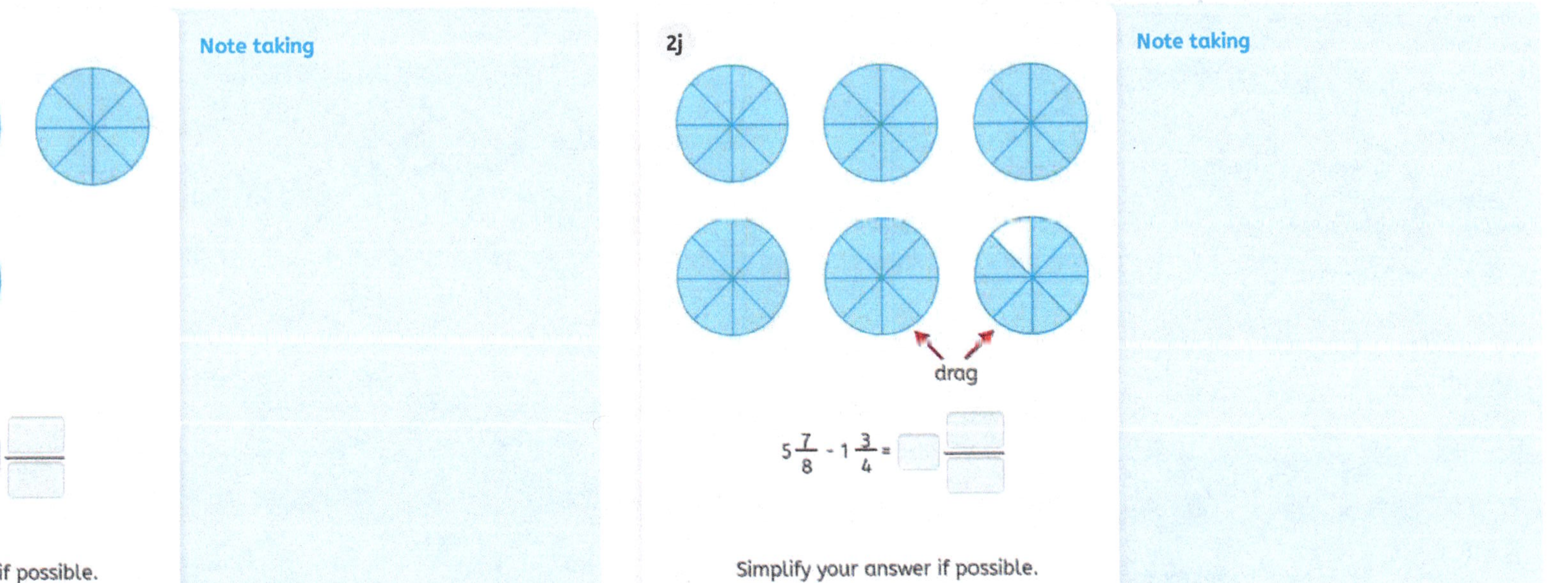

2i

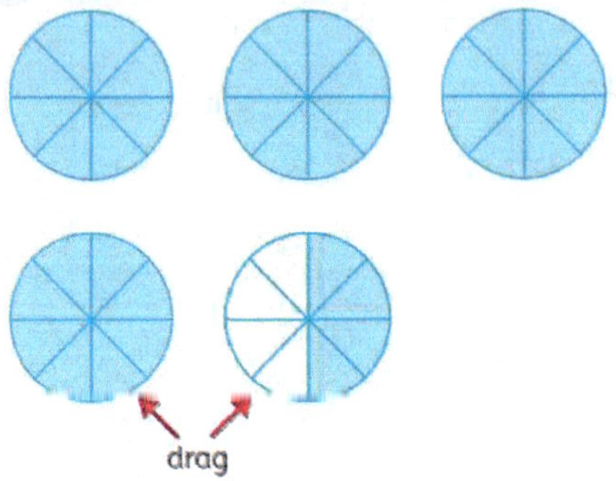

$4\frac{1}{2} - 1\frac{3}{8} =$ ☐ $\frac{☐}{☐}$

Simplify your answer if possible.

Note taking

2j

$5\frac{7}{8} - 1\frac{3}{4} =$ ☐ $\frac{☐}{☐}$

Simplify your answer if possible.

Note taking

2k Hint

$2\frac{3}{4} - 1\frac{3}{8} =$ ☐ $\frac{☐}{☐}$

☐ $\frac{☐}{8}$ ☐ $\frac{☐}{8}$

Simplify your answer if possible.

Note taking

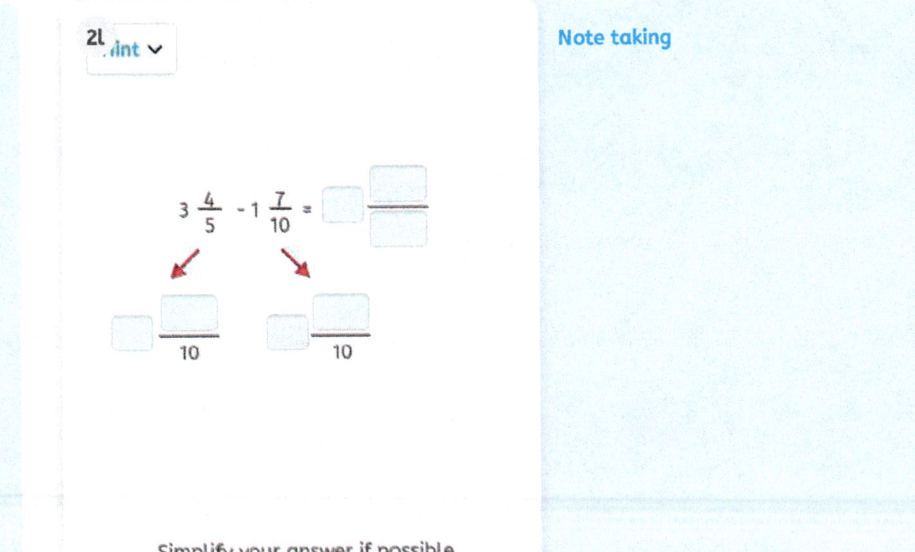

2l Hint

$3\frac{4}{5} - 1\frac{7}{10} =$ ☐ $\frac{☐}{☐}$

☐ $\frac{☐}{10}$ ☐ $\frac{☐}{10}$

Simplify your answer if possible.

Note taking

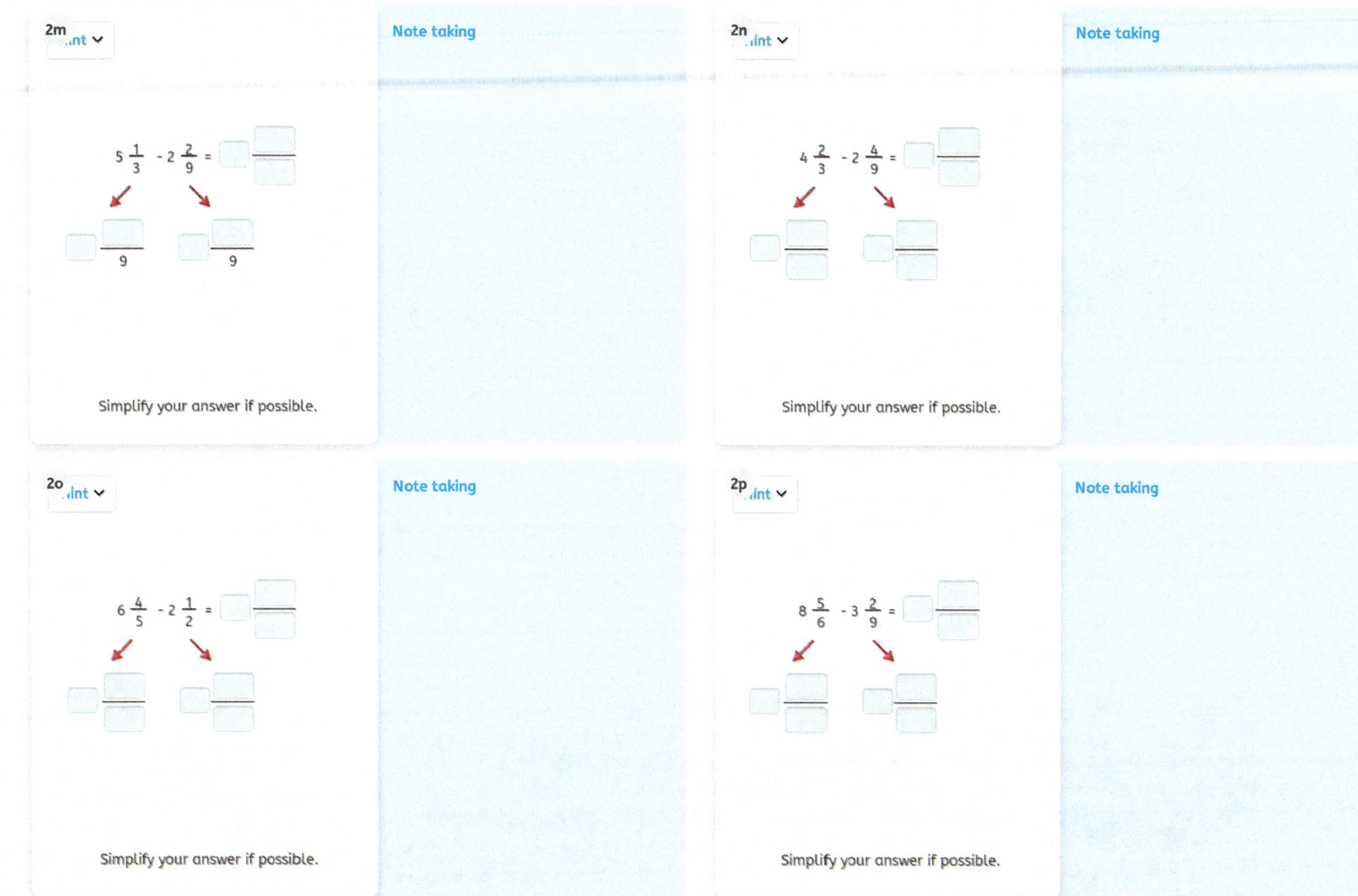

2m Hint

$5\frac{1}{3} - 2\frac{2}{9} =$

$\frac{\ }{9}$ $\frac{\ }{9}$

Simplify your answer if possible.

Note taking

2n Hint

$4\frac{2}{3} - 2\frac{4}{9} =$

Simplify your answer if possible.

Note taking

2o Hint

$6\frac{4}{5} - 2\frac{1}{2} =$

Simplify your answer if possible.

Note taking

2p Hint

$8\frac{5}{6} - 3\frac{2}{9} =$

Simplify your answer if possible.

Note taking

2q From 11 to 12 o'clock, restaurant "Adagio" served a total of 3 $\frac{1}{5}$ pots of chicken soup and 2 $\frac{3}{4}$ pots of onion soup.

How many pots of soup were served in all?

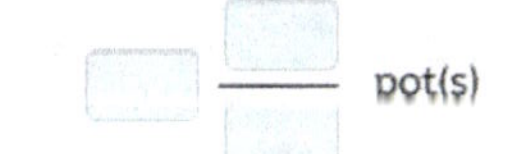

pot(s)

Simplify your answer if possible.

Note taking

2r Rebecca ran 12 $\frac{3}{4}$ miles on Friday.

She ran 10 $\frac{2}{3}$ miles on Saturday.

How much further did Rebecca run on Friday?

mile(s)

Simplify your answer if possible.

Note taking

1a

Do you remember?

$5\frac{1}{2} - 2\frac{1}{8} = \square\frac{\square}{\square}$

Note taking

1b

Do you remember?

$1\frac{1}{3} + 5\frac{1}{9} = \square\frac{\square}{\square}$

Note taking

1g

Susan eats $1\frac{1}{3}$ pancake.

Cathy eats $1\frac{1}{2}$ pancake.

Steve eats $1\frac{2}{3}$ pancake.

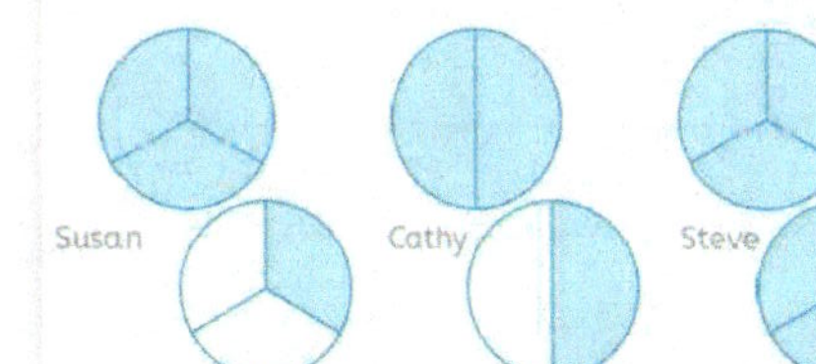
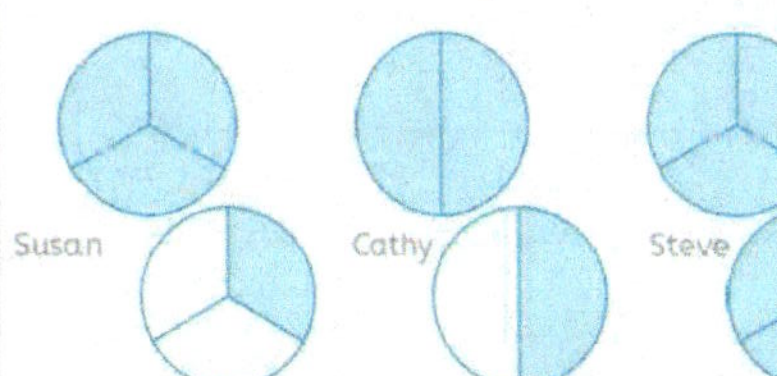

How many pancakes did they eat altogether?

Hint

They ate $\square\frac{\square}{\square}$ pancakes.

Note taking

1h

Susan buys **ONE** pizza and shares it with Cathy and Steve.

Susan eats $\frac{3}{8}$, Cathy eats $\frac{1}{4}$,

and Steve eats $\frac{1}{8}$ of the pizza.

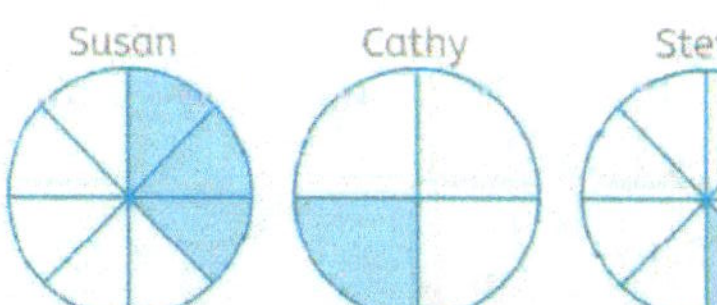

How much pizza did they eat together? Simplify the fraction.

Hint

They ate $\frac{?}{?}$ of the pizza.

Note taking

1i

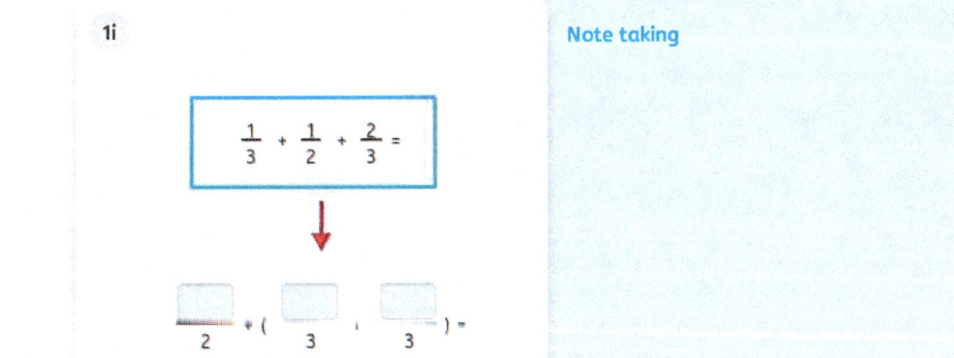

$\frac{1}{3} + \frac{1}{2} + \frac{2}{3} =$

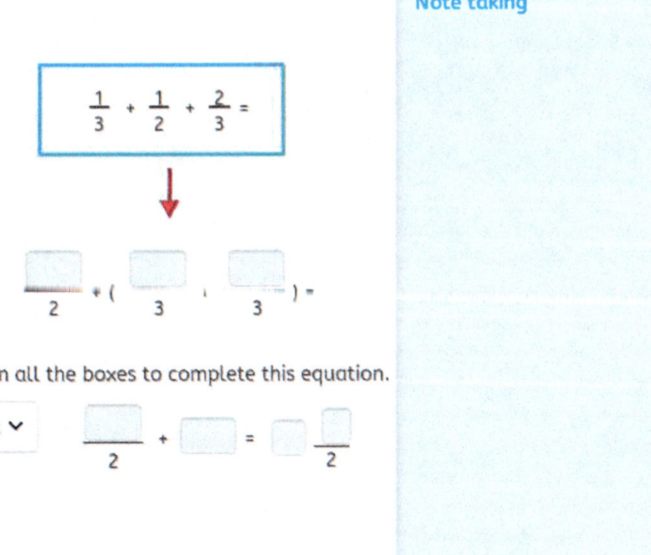

$\frac{\square}{2} + \left(\frac{\square}{3} + \frac{\square}{3} \right) =$

Fill in all the boxes to complete this equation.

Hint

$\frac{\square}{2} + \square = \square\frac{\square}{2}$

Note taking

1j

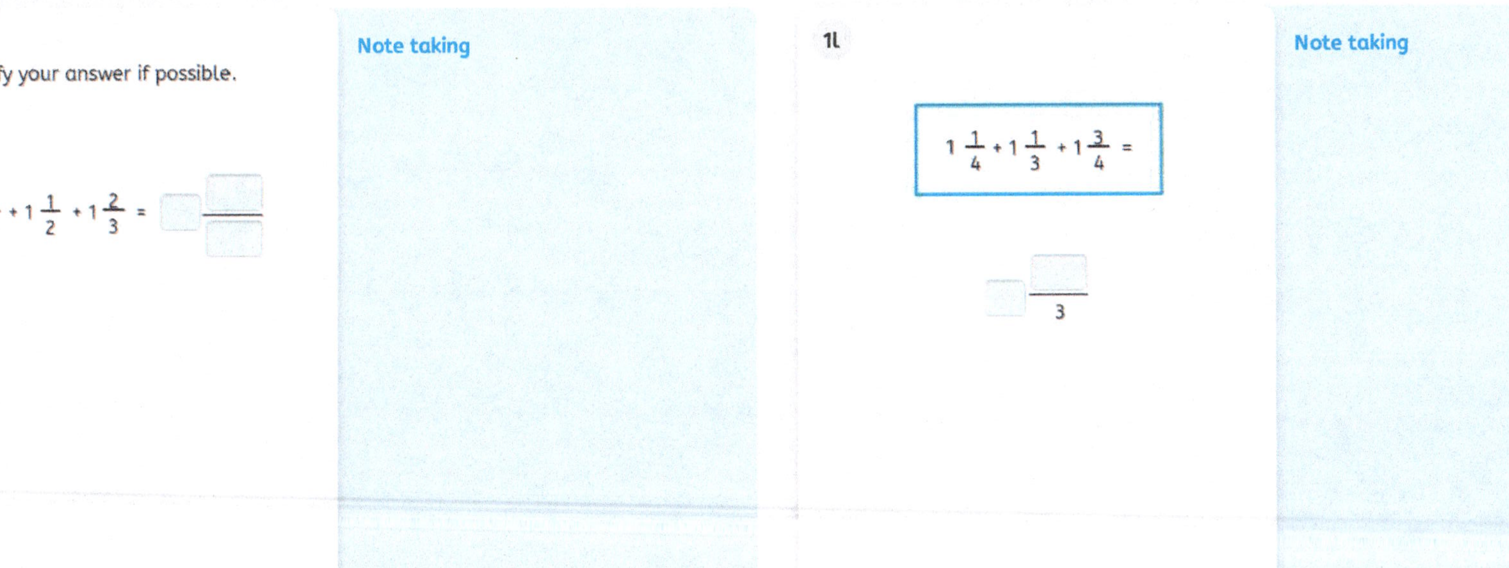

$1\frac{3}{7} + 2\frac{1}{3} + 1\frac{1}{7} =$

$\square\frac{\square}{3} + \square\frac{\square}{7} =$

$\square\frac{\square}{21}$

Note taking

1k

Simplify your answer if possible.

$2\frac{1}{3} + 1\frac{1}{2} + 1\frac{2}{3} = \square\frac{\square}{\square}$

Hint

Note taking

1l

$1\frac{1}{4} + 1\frac{1}{3} + 1\frac{3}{4} =$

$\square\frac{\square}{3}$

Note taking

2a

$\frac{1}{6} + \frac{2}{5} + \frac{1}{5} =$

$\frac{\square}{6} + \left(\frac{\square}{5} + \frac{\square}{5}\right) =$

$\frac{\square}{30}$

Order and group the fractions with like denominators.

Note taking

2b

$\frac{3}{8} + \frac{2}{3} + \frac{5}{8} =$

$\frac{\square}{3} + \left(\frac{\square}{8} + \frac{\square}{8}\right) =$

$\square\frac{\square}{\square}$

Order and group the fractions with like denominators.

Note taking

2c

$1\frac{1}{5} + 2\frac{2}{9} + 3\frac{2}{9} =$

$\square\frac{\square}{5} + \left(\square\frac{\square}{9} + \square\frac{\square}{9}\right) =$

$\square\frac{\square}{\square}$

Order and group the fractions with like denominators.

Note taking

2d

$4\frac{1}{4} + 2\frac{3}{8} + 2\frac{2}{8} =$

$\square\frac{\square}{4} + \left(\square\frac{\square}{8} + \square\frac{\square}{8}\right) =$

$\square\frac{\square}{\square}$

Order and group the fractions with like denominators.

Note taking

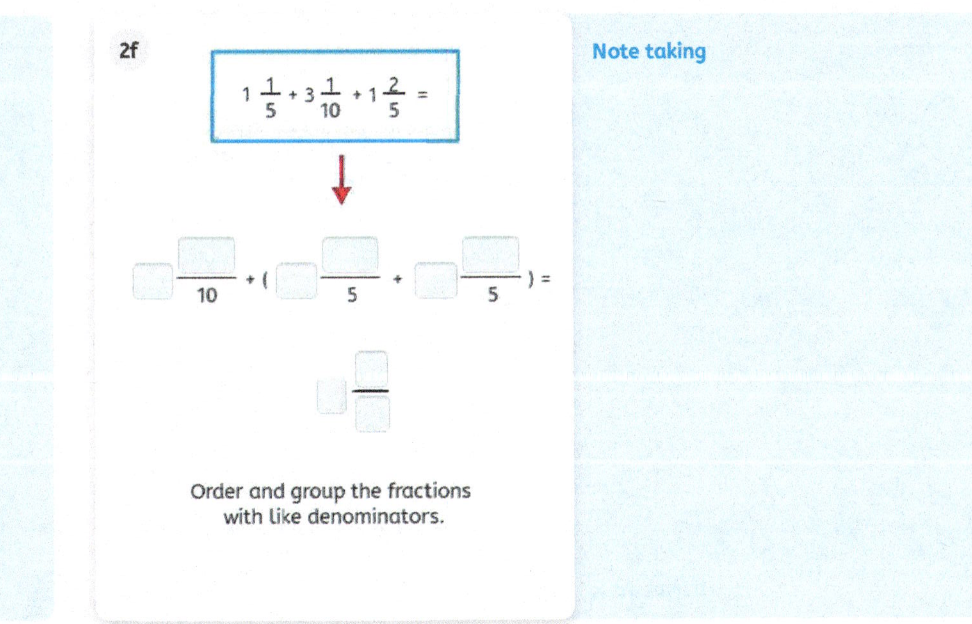

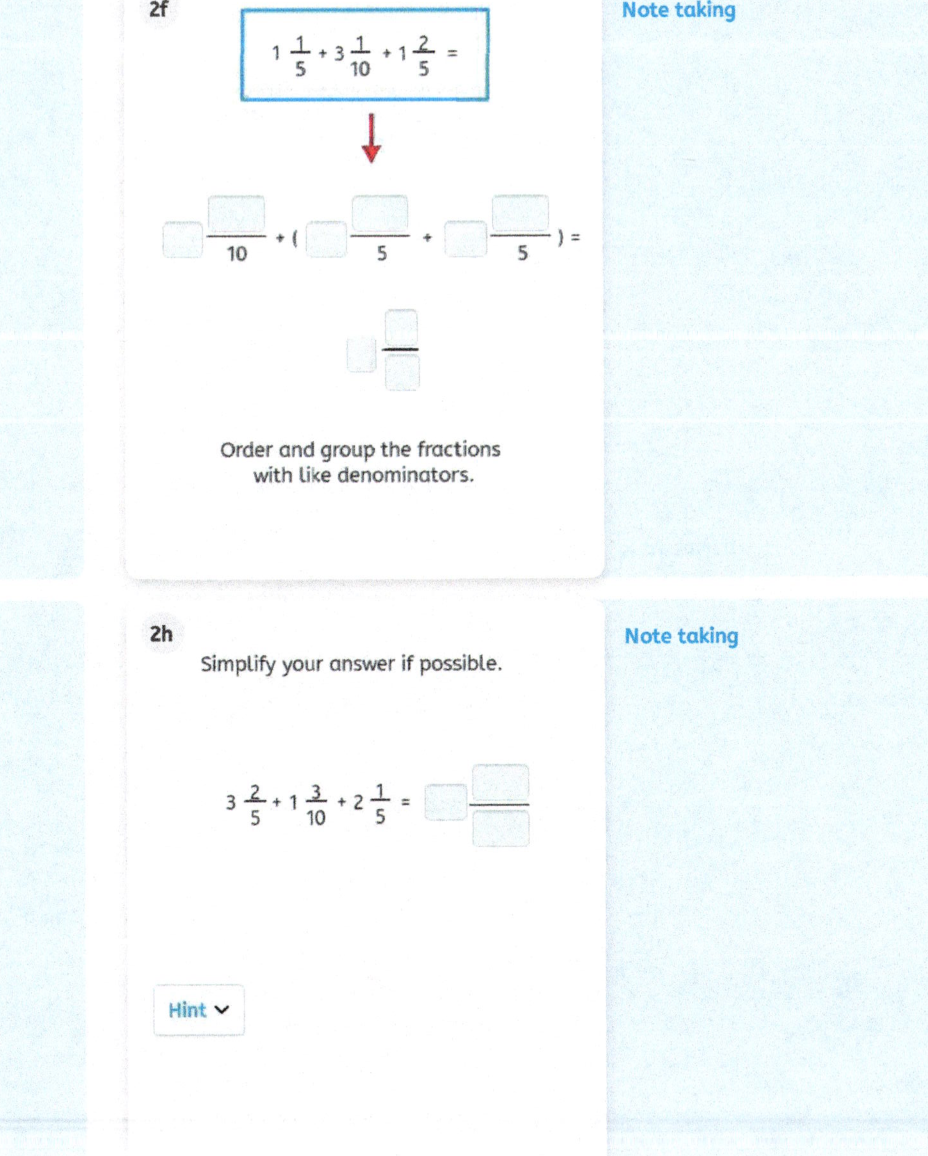

2e

$3\frac{1}{4} + 1\frac{1}{3} + 2\frac{3}{4} =$

$\square\frac{\square}{3} + (\square\frac{\square}{4} + \square\frac{\square}{4}) =$

$\square\frac{\square}{\square}$

Order and group the fractions with like denominators.

Note taking

2f

$1\frac{1}{5} + 3\frac{1}{10} + 1\frac{2}{5} =$

$\square\frac{\square}{10} + (\square\frac{\square}{5} + \square\frac{\square}{5}) =$

$\square\frac{\square}{\square}$

Order and group the fractions with like denominators.

Note taking

2g

$1\frac{2}{12} + 3\frac{1}{3} + 4\frac{5}{12} =$

$\square\frac{\square}{3} + (\square\frac{\square}{12} + \square\frac{\square}{12}) =$

$\square\frac{\square}{\square}$

Order and group the fractions with like denominators.

Note taking

2h

Simplify your answer if possible.

$3\frac{2}{5} + 1\frac{3}{10} + 2\frac{1}{5} = \square\frac{\square}{\square}$

Hint

Note taking

2i

Simplify your answer if possible.

$2\frac{2}{5} + 3\frac{3}{10} + 1\frac{3}{10} = \square$

Hint

Note taking

2j

Simplify your answer if possible.

$1\frac{1}{3} + 1\frac{1}{6} + 1\frac{5}{6} = \square\frac{\square}{\square}$

Hint

Note taking

2k

Simplify your answer if possible.

$3\frac{2}{3} + 4\frac{1}{9} + 2\frac{2}{3} = \square\frac{\square}{\square}$

Hint

Note taking

2l

Simplify your answer if possible.

$2\frac{4}{9} + 1\frac{2}{3} + 5\frac{5}{9} = \square\frac{\square}{\square}$

Hint

Note taking

2m

Steven walked for 3 days.

On day one, he walked $2\frac{5}{6}$ miles.

On day two, he walked $3\frac{1}{3}$ miles.

On day three, he walked $2\frac{2}{3}$ miles.

Steven walked $\square\frac{\square}{\square}$ miles in 3 days.

Note taking

1a Do you remember?

Susan eats $1\frac{1}{3}$ of a pancake.

Cathy eats $1\frac{1}{2}$ of a pancake.

Steve eats $1\frac{2}{3}$ of a pancake.

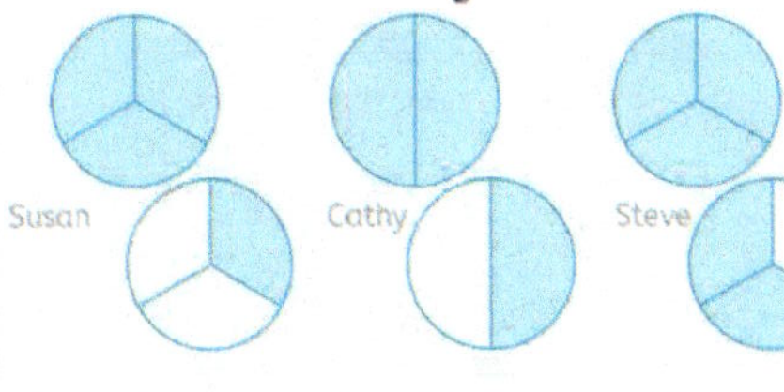

How many pancakes did they eat altogether?

They ate ☐ $\frac{☐}{☐}$ pancakes.

Note taking

1b

Do you remember?

$1\frac{3}{7} + 2\frac{1}{3} + 1\frac{1}{7} =$ ☐ $\frac{☐}{21}$

Note taking

1f

David shares 3 pizzas.

Sharon eats $\frac{3}{4}$ pizza.

Bobby eats $\frac{1}{2}$ pizza.

Ruby eats $\frac{7}{8}$ pizza.

How much pizza is left?

Explain how you would solve the problem.

Note taking

1g

David shares 3 pizzas.

Sharon eats $\frac{3}{4}$ pizza.

Bobby eats $\frac{1}{2}$ pizza.

Ruby eats $1\frac{1}{8}$ pizza.

What expression can you write to show how much pizza is left?

Note taking

1h

David shares 3 pizzas.

Sharon eats $\frac{3}{4}$ pizza.

Bobby eats $\frac{1}{2}$ pizza.

Ruby eats $\frac{7}{8}$ pizza.

How much pizza is left?

Which 2 equations model the problem?

$? + \frac{3}{4} + \frac{1}{2} + \frac{7}{8} = 3$

$? + 3 + \frac{3}{4} + \frac{1}{2} = \frac{7}{8}$

$? - \frac{3}{4} - \frac{1}{2} - \frac{7}{8} = 3$

$3 - \frac{3}{4} - \frac{1}{2} - \frac{7}{8} = ?$

Note taking

1i

David shares 3 pizzas.

Sharon eats $\frac{3}{4}$ pizza.

Bobby eats $\frac{1}{2}$ pizza.

Ruby eats $\frac{7}{8}$ pizza.

How much pizza is left?

Solve the problem.

There is $\frac{?}{?}$ pizza left.

Note taking

1j

David shares 3 pizzas.

Sharon eats $\frac{3}{4}$ pizza.

Bobby eats $\frac{1}{2}$ pizza.

Ruby eats $\frac{7}{8}$ pizza.

How much pizza is left?

David says: '*There is* $2\frac{1}{8}$ *pizza left*'.

Is David right or wrong? Explain why.

Note taking

1k

Cookie Recipe
(for 8 cookies)

- 1/2 cup butter
- 1/4 cup chocolate chips
- 4/8 cup sugar
- 3/4 cup flour

How much of each ingredient does Chris need need to make enough cookies for each of his 4 friends to have 6 cookies each?

Explain using pictures, numbers, and words.

Note taking

2a

David has 2 pizzas.

Cheryl eats $\frac{1}{4}$ pizza.

Bob eats $\frac{1}{2}$ pizza.

David eats the rest.

How much does David eat?

Explain how you would solve the problem.

Note taking

2b

Toby knitted 1 scarf.

$\frac{2}{3}$ of the scarf is red.

$\frac{1}{5}$ of the scarf is grey.

The rest of the scarf is orange.

What part is orange?

Explain how you would solve the problem.

Note taking

2c

David has 2 pizzas.

Sharol eats $\frac{1}{4}$ pizza.

Bob eats $\frac{1}{2}$ pizza.

David eats the rest.

How much does David eat?

Solve the problem.

David eats ☐ $\frac{☐}{☐}$ pizza.

Note taking

2d

Toby knitted a scarf.

$\frac{2}{3}$ of the scarf is red.

$\frac{1}{5}$ of the scarf is grey.

The rest of the scarf is orange.

What part is orange?

Solve the problem.

$\frac{?}{?}$ part is orange.

Note taking

2e

Joe has 5 boxes.

$1\frac{2}{5}$ box is filled with books.

$1\frac{1}{2}$ box is filled with CDs.

The rest of the box is filled with clothes.

Solve the problem.

$\square\frac{\square}{\square}$ box is filled with clothes.

Note taking

2f

Dan painted 4 walls.

He painted $1\frac{1}{3}$ wall white.

He painted $1\frac{5}{6}$ wall blue.

He painted the rest black.

Solve the problem.

$\frac{?}{?}$ wall is painted black.

Note taking

2g

David has 2 pizzas.

Sharol eats $\frac{1}{4}$ pizza.

Bob eats $\frac{1}{2}$ pizza.

David eats the rest.

How much does David eat?

David says: *'I ate more than 1 pizza.'*

Is David right or wrong? Explain why.

Note taking

2h

Toby knitted a scarf.

$\frac{2}{3}$ of the scarf is red.

$\frac{1}{5}$ of the scarf is grey.

The rest of the scarf is orange.

What part is orange?

Toby says: '*Most of the scarf is orange*'.

Is Toby right or wrong? Explain why.

Note taking

2i

Joe has 5 boxes.

$1\frac{2}{5}$ box is filled with books.

$1\frac{1}{2}$ box is filled with CDs.

The rest is filled with clothes.

Joe says: 'More than 3 boxes are *filled with clothes*.'

Is Joe right or wrong? Explain why.

Note taking

2j

Dan painted 4 walls.

He painted $1\frac{1}{3}$ wall white.

He painted $1\frac{5}{6}$ wall blue.

He painted the rest black.

Dan says: '*I painted less than 1 wall black*.'

Is Dan right or wrong? Explain why.

Note taking

2k

At the end of a science experiment 4 groups had leftover water in their beakers.

See below:

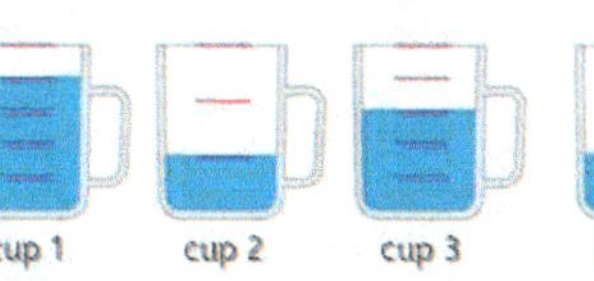

What are some ways to combine cups without overflowing?

Write two combinations down below:

Note taking

1a

Do you remember?

Compare the fractions. Which one is greater?

$\frac{2}{5}$ $\frac{2}{10}$

Note taking

1b

Do you remember?

Which fraction is greater?

$\frac{2}{6}$ $\frac{3}{8}$

Note taking

1f D the description of each product to its picture.

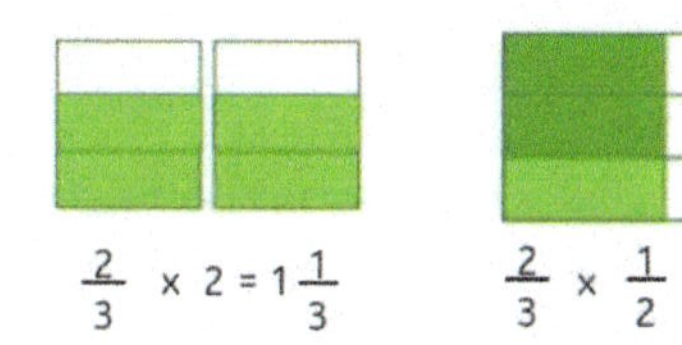

$\frac{2}{3} \times 2 = 1\frac{1}{3}$

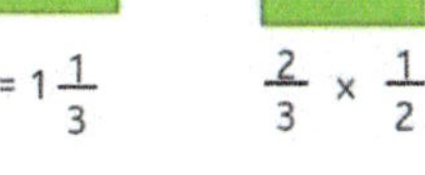

$\frac{2}{3} \times \frac{1}{2} = \frac{1}{3}$

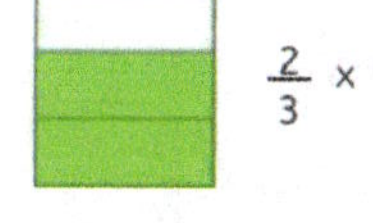

$\frac{2}{3} \times 1 = \frac{2}{3}$

product smaller than both factors

product smaller than one factor

Note taking

1g

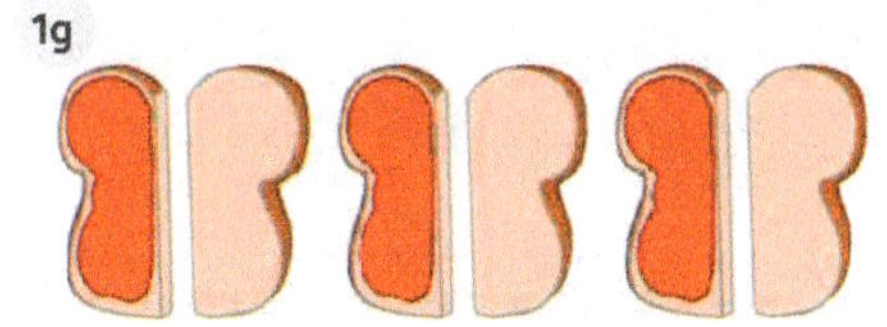

$\frac{1}{2}$ of 3 sandwiches are covered in jelly.

$\frac{1}{2} \times 3 = 1\frac{1}{2}$

Which statement fits the product?

- The product is greater than both factors.
- The product is less than one factor.
- The product is less than both factors.

Note taking

1h

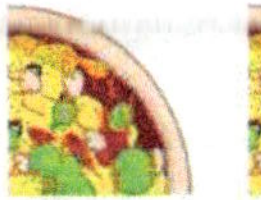

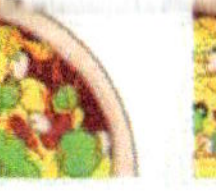

3 pieces of $\frac{1}{4}$ pizza.

$3 \times \frac{1}{4} = \frac{3}{4}$

Which statement is correct?

- The product is greater than both factors.
- The product is less than both factors.
- The product less than one factor.

Note taking

1i

$1 \times \frac{1}{3} = \frac{1}{3}$ $2 \times \frac{1}{3} = \frac{2}{3}$ $\frac{1}{2} \times \frac{1}{3} = \frac{1}{6}$

Drag the right models to their products.

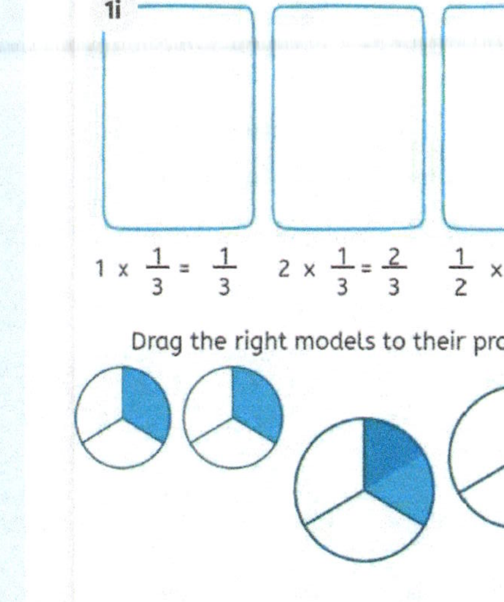

Note taking

1j

$1 \times \frac{1}{3} = \frac{1}{3}$ $2 \times \frac{1}{3} = \frac{2}{3}$ $\frac{1}{2} \times \frac{1}{3} = \frac{1}{6}$

What happens to the size of a product when you multiply one or more fractions?

less/greater

The product is ______ than one or more of the factors.

Note taking

1k Drag the correct phrases into the sentence.

$\frac{3}{4} \times \frac{1}{2} = \frac{3}{8}$

less

greater

The product is ______ than ______ of the factors.

one both

Note taking

1l Drag the correct phrases into the sentence.

$\frac{4}{5} \times 1 = \frac{4}{5}$

less

greater

The product is ________ than ________ of the factors.

one both

Note taking

1m Drag the correct phrase into the sentence.

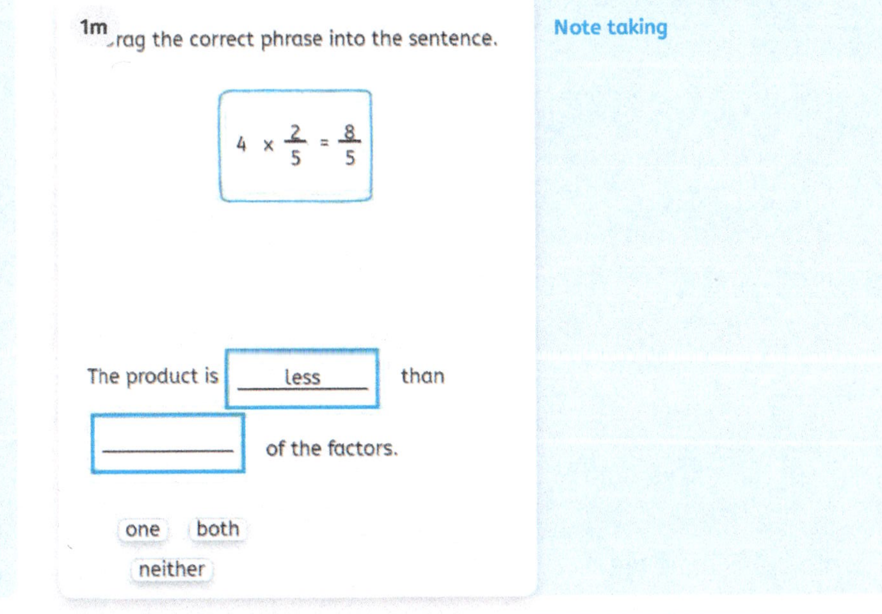

$4 \times \frac{2}{5} = \frac{8}{5}$

The product is less than ________ of the factors.

one both

neither

Note taking

1n Drag the correct phrase into the sentence.

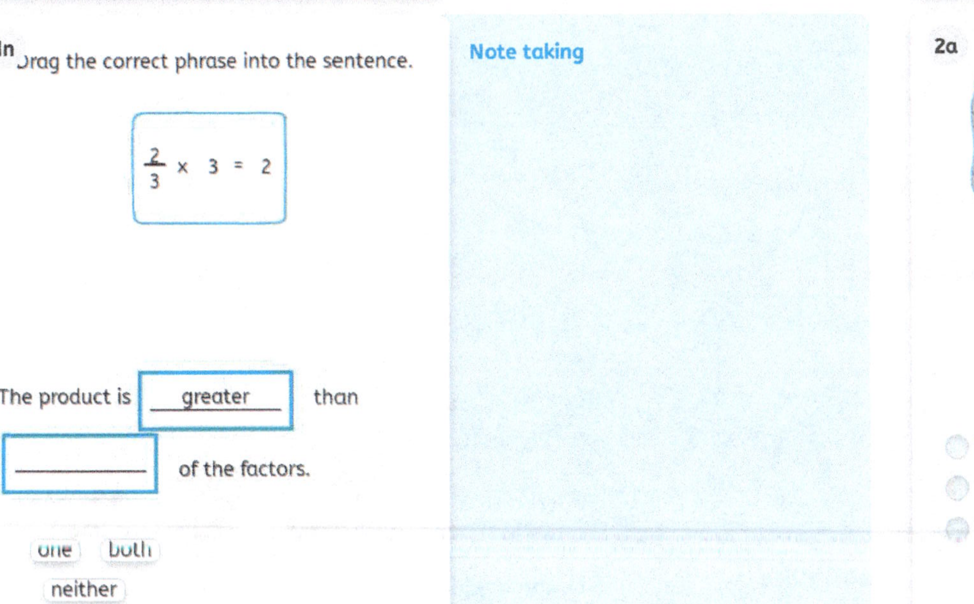

$\frac{2}{3} \times 3 = 2$

The product is greater than ________ of the factors.

one both

neither

Note taking

2a

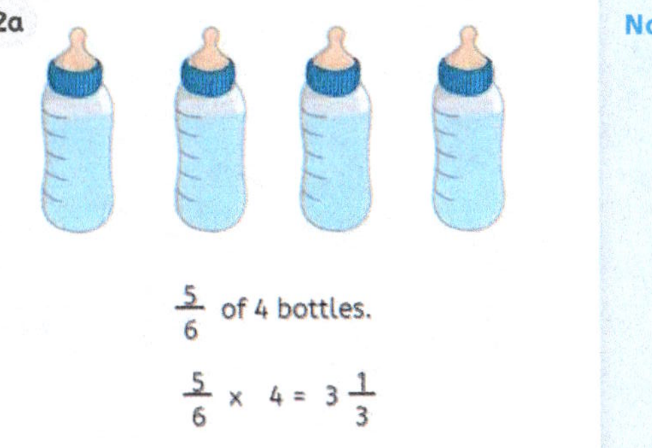

$\frac{5}{6}$ of 4 bottles.

$\frac{5}{6} \times 4 = 3\frac{1}{3}$

Which statement is correct?

- The product is greater than both factors.
- The product is less than one factor.
- The product is less than both factors.

Note taking

2b

5 pieces of $\frac{3}{4}$ cake.

$5 \times \frac{3}{4} = 3\frac{3}{4}$

Which statement is correct?

- ◯ The product is less than one factor.
- ◯ The product is less than both factors.
- ◯ The product is greater than both factors.

Note taking

2c

$\frac{3}{5}$ of a tree trunk

$\frac{3}{5} \times 1 = \frac{3}{5}$

Which statement fits the product?

- ◯ The product is less than one factor.
- ◯ The product is less than both factors.
- ◯ The product is greater than both factors.

Note taking

2d

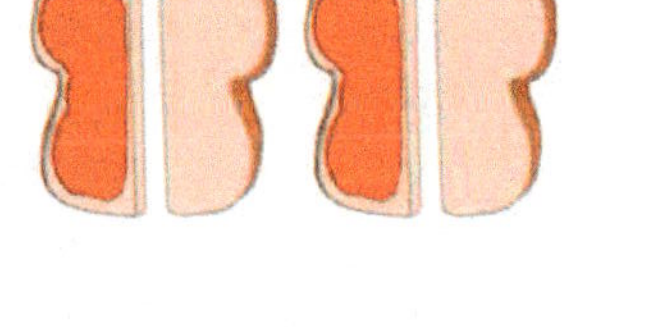

$\frac{1}{2}$ of 2 sandwiches are covered in jelly.

$\frac{1}{2} \times 2 = 1$

Which statement fits the product?

- ◯ The product is greater than both factors.
- ◯ The product is less than one factor.
- ◯ The product is less than both factors.

Note taking

2e

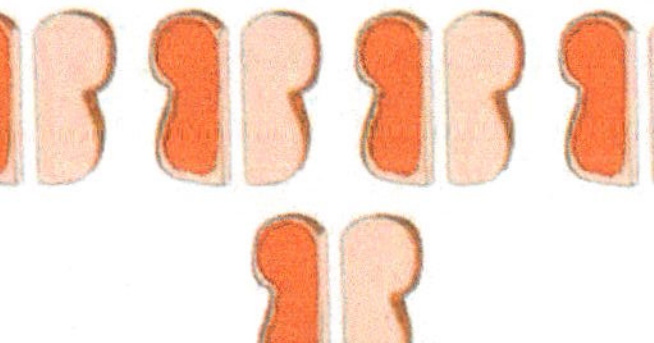

$\frac{1}{2}$ of 5 sandwiches are covered in jelly.

$\frac{1}{2} \times 5 = 2\frac{1}{2}$

Which statement fits the product?

- ◯ The product is less than one factor.
- ◯ The product is greater than both factors.
- ◯ The product is less than both factors.

Note taking

2f Drag the correct phrase into the sentence.

$7 \times \frac{1}{4} = 1\frac{3}{4}$

The product is [less] than [______] of the factors.

one | both | neither

Note taking

2g Drag the correct phrase into the sentence.

$\frac{3}{4} \times 4 = 3$

The product is [greater] than [______] of the factors.

one | both | neither

Note taking

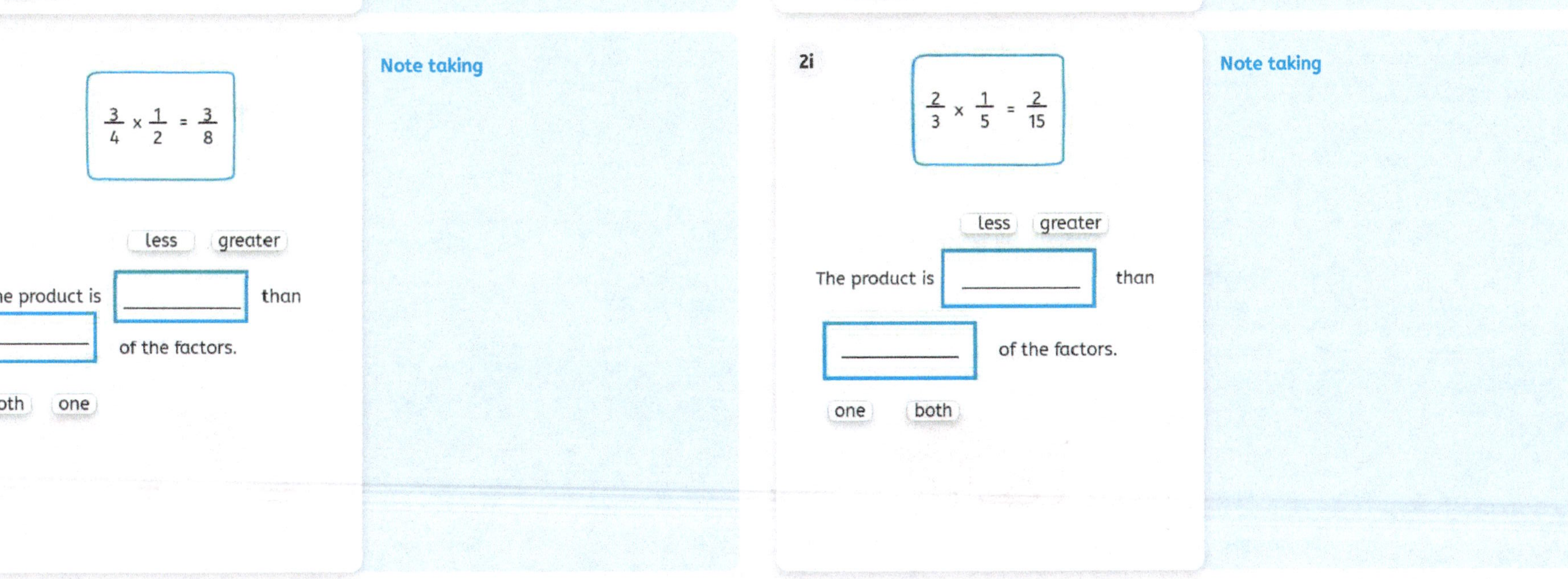

2h

$\frac{3}{4} \times \frac{1}{2} = \frac{3}{8}$

less | greater

The product is [______] than [______] of the factors.

both | one

Note taking

2i

$\frac{2}{3} \times \frac{1}{5} = \frac{2}{15}$

less | greater

The product is [______] than [______] of the factors.

one | both

Note taking

2j

$\frac{7}{8} \times 1 = \frac{7}{8}$

less greater

The product is ______ than ______ of the factors.

one both

Note taking

2k

$\frac{5}{6} \times \frac{5}{6} = \frac{25}{36}$

less greater

The product is ______ than ______ of the factors.

one both

Note taking

2l

$\frac{9}{10} \times 1 = \frac{9}{10}$

less greater

The product is ______ than ______ of the factors.

one both

Note taking

2m

5 pieces of $\frac{1}{4}$ pizza.

$5 \times \frac{1}{4} = 1\frac{1}{4}$

Which statement is correct?

- The product is greater than both factors.
- The product is less than both factors.
- The product is less than one factor.

Note taking

2n

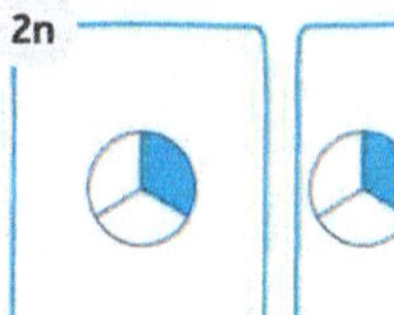

$1 \times \frac{1}{3} = \frac{1}{3}$

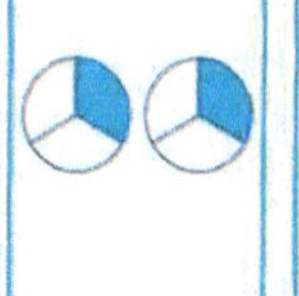

$2 \times \frac{1}{3} = \frac{2}{3}$

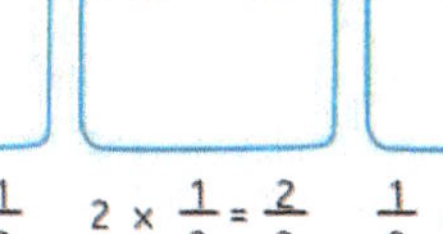

$\frac{1}{2} \times \frac{1}{3} = \frac{1}{6}$

Which product is less than both factors?

Note taking

1a

Do You Remember?

Drag the right words into the sentence.

$\frac{5}{4} \times 1$

The product will be ________ 1.

less than

greater than

equal to

Note taking

1b

Game: Rearrange!

You have 18 strawberries, and you eat $\frac{2}{3}$ of them.

Extra

How can you arrange the strawberries to find $\frac{2}{3}$?

You divide them in ☐ equal groups.

Each group represents $\frac{☐}{☐}$.

Note taking

1j

$6 \times \frac{1}{7} = \frac{☐}{☐}$

Note taking

1k

Hint

Eric has 15 marbles.

He gives $\frac{1}{5}$ of them to Stephen.

$\frac{1}{5}$ of 15 = ☐ , or $\frac{1}{5} \times 15 =$ ☐

So, Eric gives ☐ marbles to Stephen.

Note taking

1l

Eric has 24 marbles.

He gives $\frac{3}{4}$ of them to Stephen.

$\frac{3}{4}$ of 24 = ☐ , or $\frac{3}{4} \times 24 =$ ☐

So, Eric gives ☐ marbles to Stephen.

Note taking

1m

Calculate $\frac{1}{3}$ of 9, or $\frac{1}{3} \times 9$.

Draw the model below on a piece of paper. Circle the rows that you need to solve the problem.

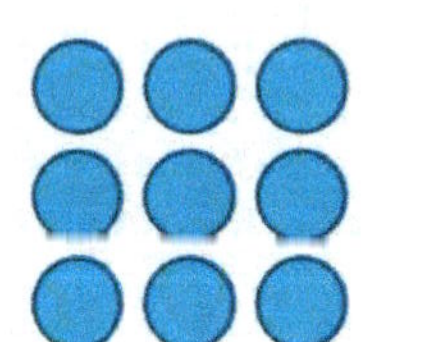

$\frac{1}{3} \times 9 =$ ☐

Note taking

1n

$4 \times \frac{2}{9} = \frac{☐}{☐}$

Note taking

1o

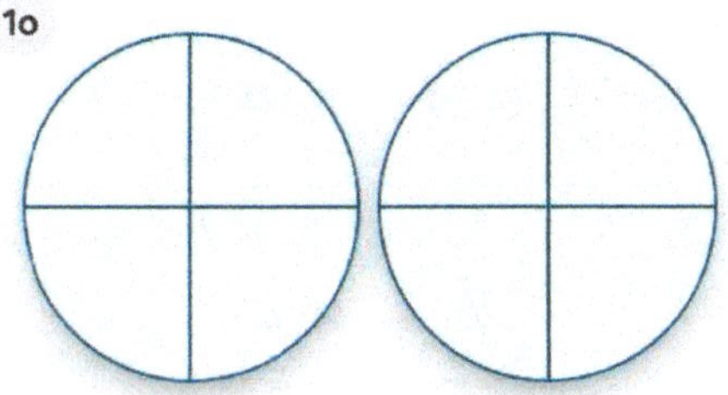

$7 \times \frac{1}{4} = \frac{☐}{4} = 1\frac{☐}{4}$

Note taking

1p

$2 \times \frac{2}{5} = \frac{\square}{\square}$

Note taking

1q

$2 \times \frac{4}{9} = \frac{\square}{\square}$

Note taking

1r

$11 \times \frac{1}{6} = \frac{\square}{\square} = \square \frac{\square}{\square}$

Note taking

2a

Eric has 12 marbles.

He gives $\frac{3}{4}$ of them to Stephen.

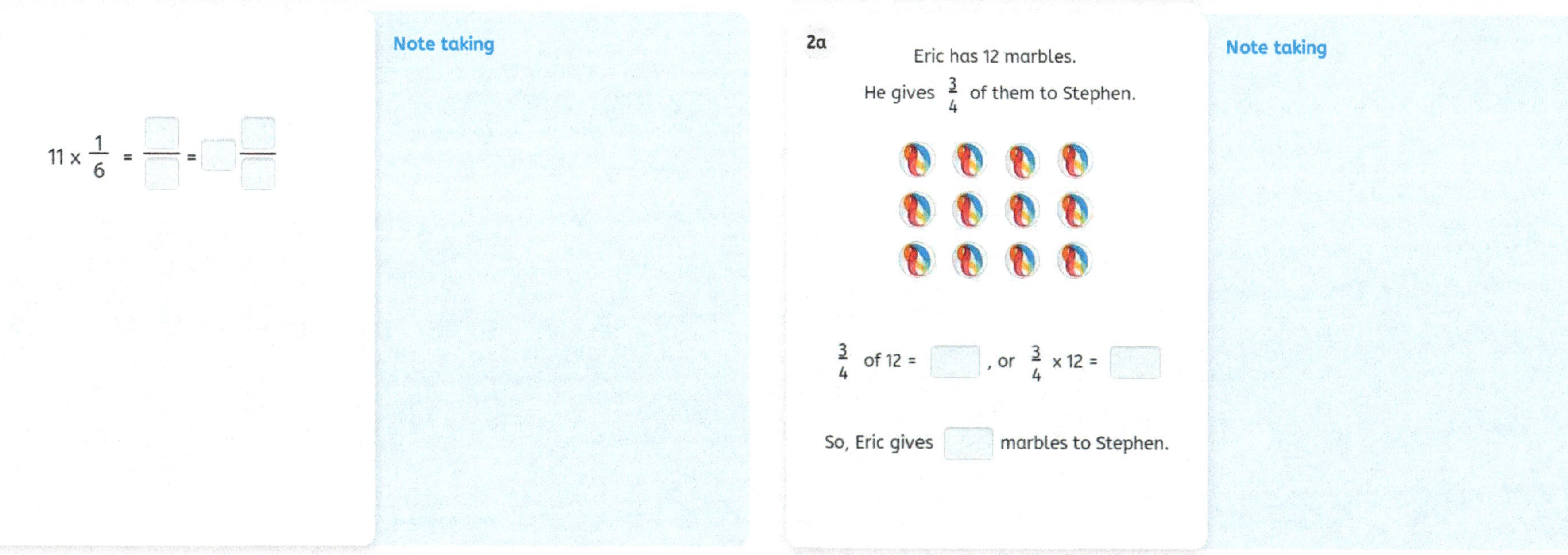

$\frac{3}{4}$ of 12 = ☐, or $\frac{3}{4} \times 12 =$ ☐

So, Eric gives ☐ marbles to Stephen.

Note taking

2b

Calculate $\frac{2}{3}$ of 21, or $\frac{2}{3} \times 21$.

Draw the model below on a piece of paper. Circle the rows that you need to solve the problem.

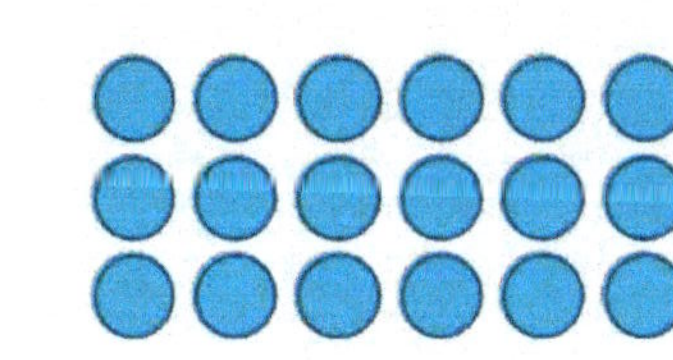

$\frac{2}{3} \times 21 =$ ☐

Note taking

2c

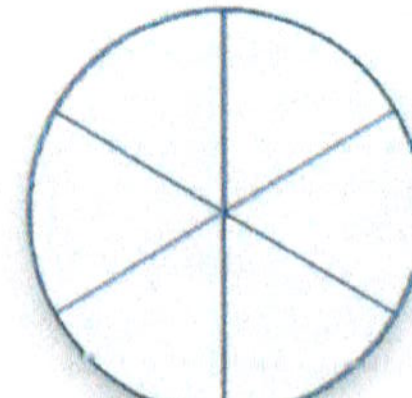

$5 \times \frac{1}{6} = \frac{☐}{☐}$

Note taking

2d

$2 \times \frac{2}{9} = \frac{☐}{☐}$

Note taking

2e

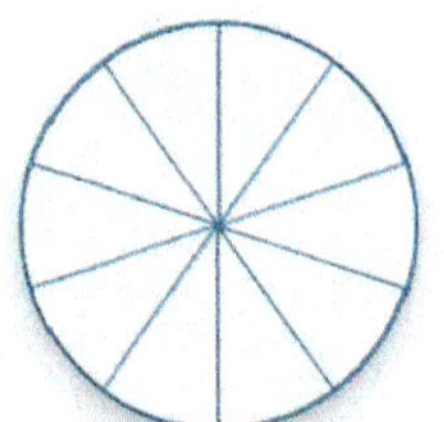

$3 \times \frac{3}{10} = \frac{☐}{☐}$

Note taking

2f

$5 \times \frac{1}{3} = \frac{\square}{3} = 1\frac{\square}{3}$

Note taking

2g

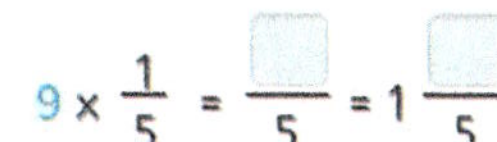

$9 \times \frac{1}{5} = \frac{\square}{5} = 1\frac{\square}{5}$

Note taking

2h

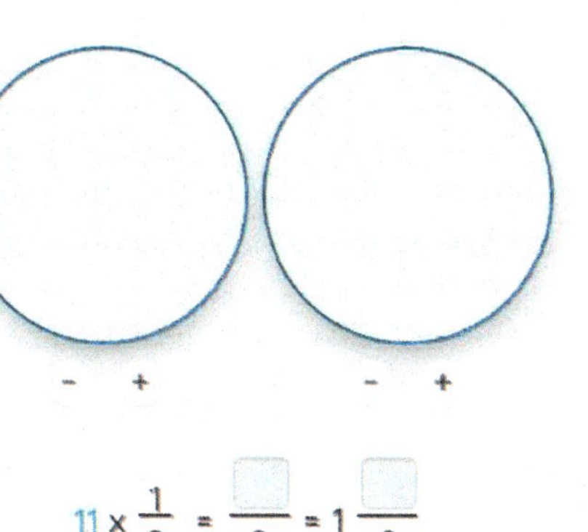

$11 \times \frac{1}{8} = \frac{\square}{8} = 1\frac{\square}{8}$

Note taking

2i

$3 \times \frac{2}{7} = \frac{\square}{\square}$

Note taking

2j

$2 \times \frac{4}{9} = \frac{\square}{\square}$

Note taking

2k

$8 \times \frac{1}{7} = \frac{\square}{\square} = \square\frac{\square}{\square}$

Note taking

2l

$5 \times \frac{3}{8} = \frac{\square}{\square} = \square\frac{\square}{\square}$

Note taking

1a

Do You Remember?

$3 \times \frac{1}{4} = \frac{?}{?}$

Note taking

1b

Do You Remember?

$2 \times \frac{3}{7} = \frac{?}{?}$

Note taking

1g

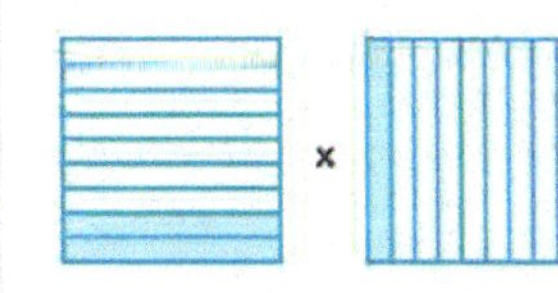

$\frac{2}{9} \times \frac{1}{9} = \frac{?}{81}$

Note taking

1h

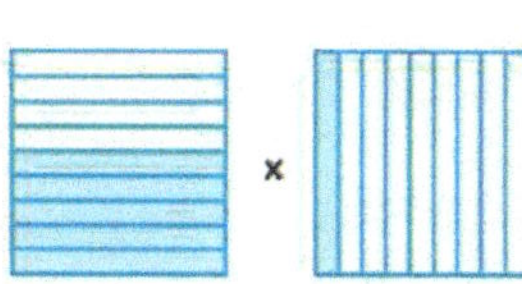

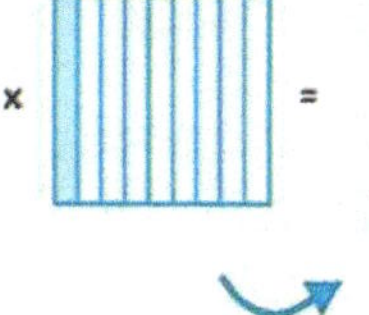

$\frac{5}{9} \times \frac{1}{9} = \frac{?}{81}$

Note taking

1i

$\frac{5}{9} \times \frac{1}{2}$

× =

Click on the correct product.

$\frac{5}{18}$ $\frac{2}{18}$ $\frac{5}{2}$

Note taking

1j

$\frac{1}{2} \times \frac{2}{4}$

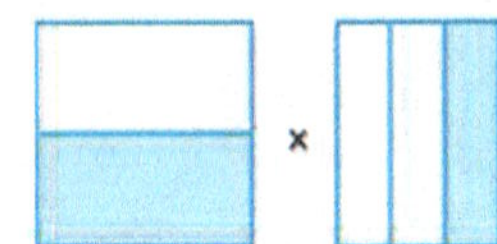

× =

Click on the correct product.

$\frac{2}{6}$ $\frac{1}{8}$ $\frac{2}{8}$

Note taking

1k

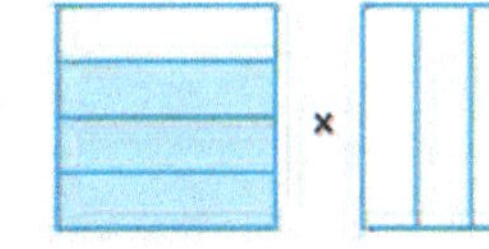

× =

$\frac{3}{4} \times \frac{1}{4} = \frac{?}{?}$

Note taking

1L

× =

$\frac{2}{3} \times \frac{2}{3} = \frac{?}{?}$

Note taking

2a

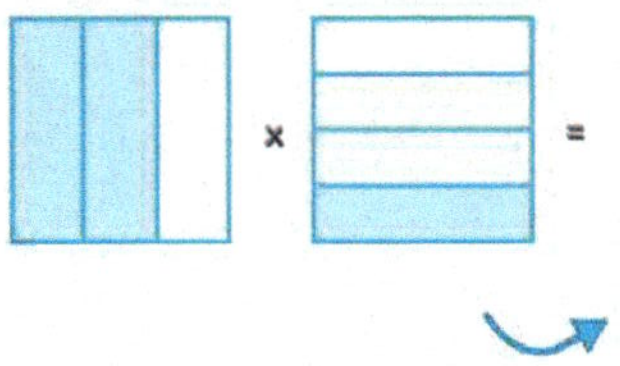

$\frac{2}{3} \times \frac{1}{4} = \frac{?}{?}$

Note taking

2b

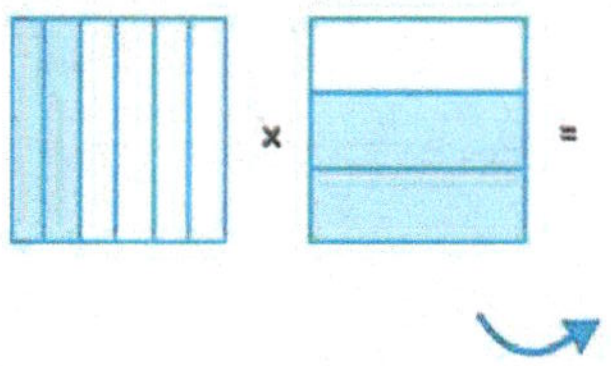

$\frac{2}{6} \times \frac{2}{3} = \frac{?}{?}$

Note taking

2c

$\frac{3}{9} \times \frac{1}{9}$

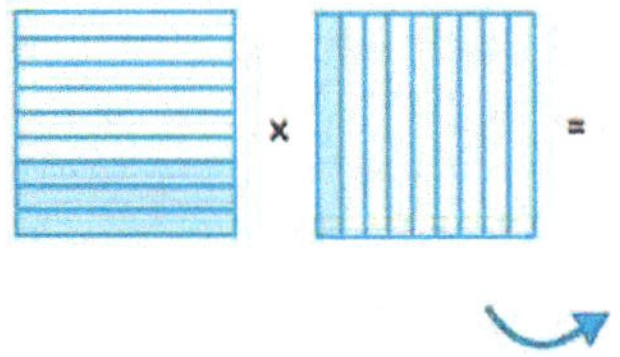

Click on the correct product.

$\frac{1}{81}$ $\frac{1}{9}$ $\frac{3}{81}$

Note taking

2d

$\frac{1}{2} \times \frac{3}{4}$

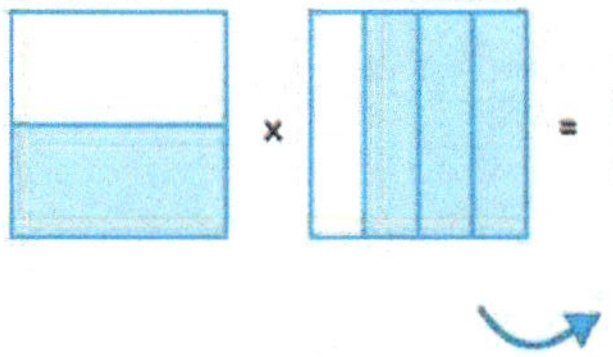

Click on the correct product.

$\frac{3}{8}$ $\frac{3}{6}$ $\frac{1}{8}$

Note taking

2e

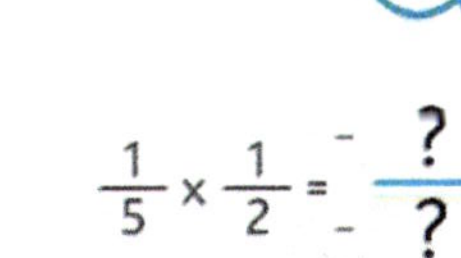

$\frac{1}{5} \times \frac{1}{2} = \frac{?}{?}$

Note taking

2f

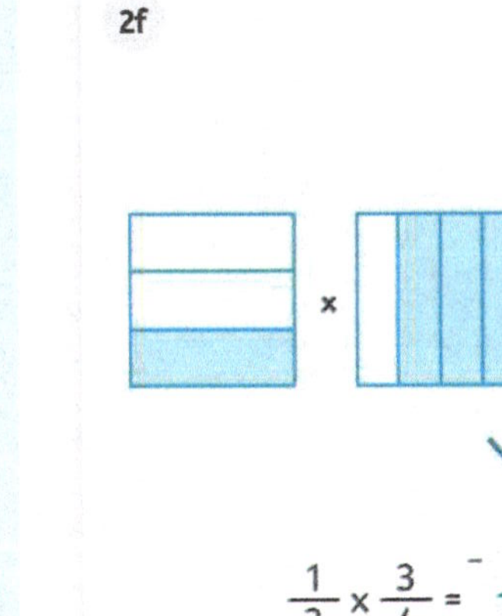

$\frac{1}{3} \times \frac{3}{4} = \frac{?}{?}$

Note taking

2g

$\frac{3}{4} \times \frac{1}{4} = \frac{?}{?}$

Note taking

2h

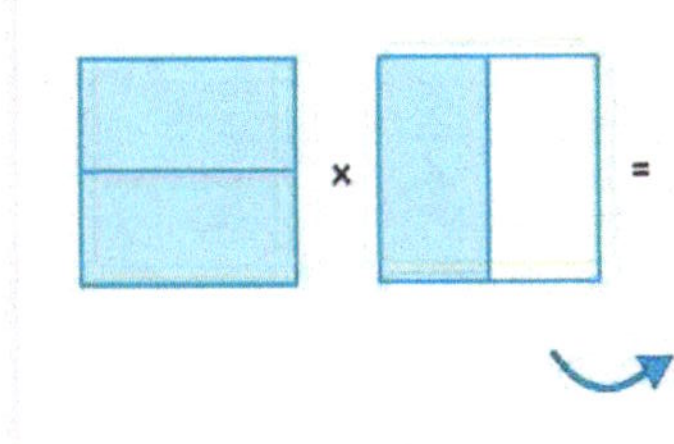

$\frac{2}{2} \times \frac{1}{2} = \frac{?}{?}$

Note taking

2i

× =

$$\frac{3}{5} \times \frac{1}{2} = \frac{?}{?}$$

Note taking

1a

Do You Remember?

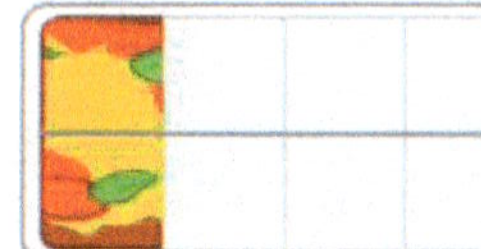

There's still $\frac{1}{4}$ part of lasagna.

Ryan eats $\frac{1}{2}$ part of that.

What part of lasagna does Ryan eat?

$\frac{1}{2} \times \frac{1}{4} = \frac{?}{?}$ part lasagna

Note taking

1b

Do You Remember?

There is $\frac{3}{4}$ of a fruit tart.

He eats $\frac{1}{3}$ part of that.

How much does he eat?

$\frac{1}{3} \times \frac{3}{4} = \frac{?}{?}$

Note taking

1k

$\frac{1}{3} \times \frac{1}{4} = \frac{?}{?}$

Note taking

1l

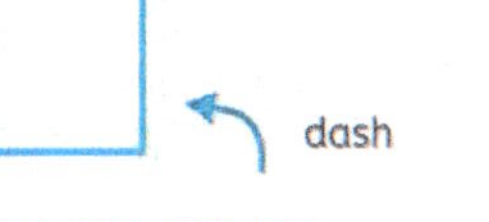

dash

$\frac{1}{2} \times \frac{1}{3} = \frac{?}{?}$

Note taking

1m

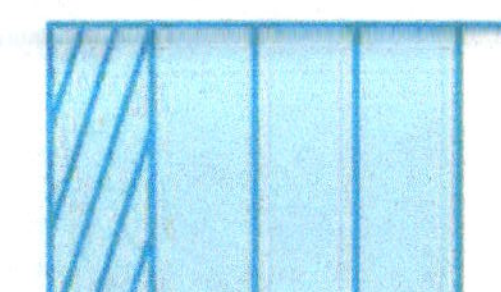

$\frac{1}{4} \times \frac{4}{5} = \frac{?}{?}$

Note taking

1n

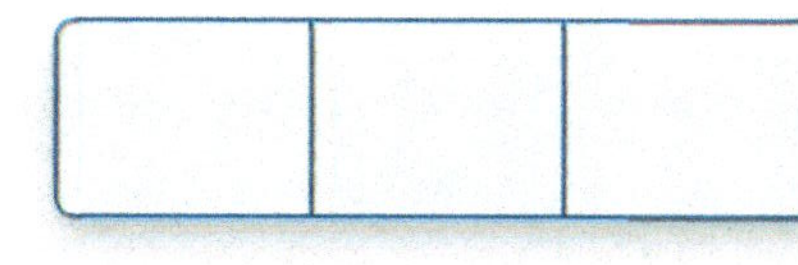

$\frac{1}{2} \times \frac{2}{3} = \frac{?}{?}$

Note taking

1o

$\frac{2}{4} \times \frac{2}{3} = \frac{?}{?}$

Note taking

2a

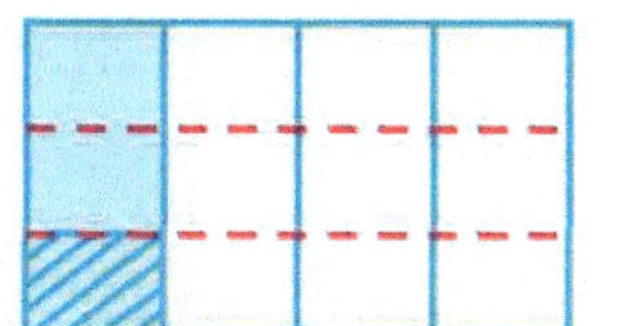

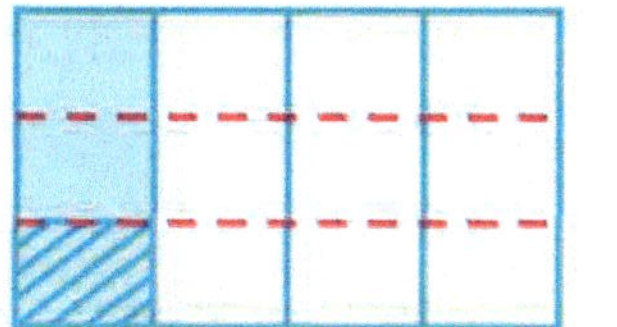

$\frac{1}{3} \times \frac{1}{4} = \frac{?}{?}$

Note taking

2b

$\frac{1}{2} \times \frac{1}{3} = \frac{?}{?}$

Note taking

2c

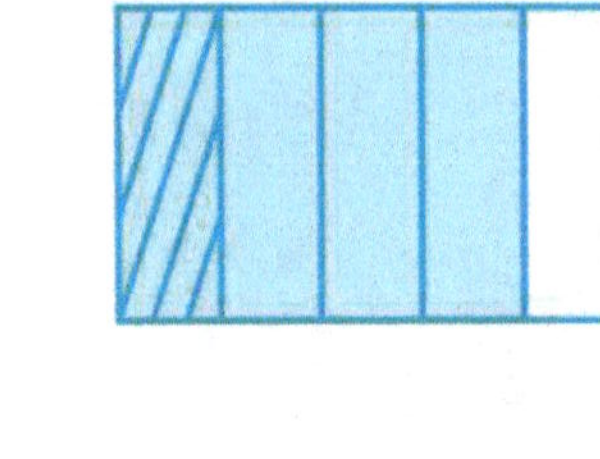

$\frac{1}{4} \times \frac{4}{5} = \frac{?}{?}$

Note taking

2d

$\frac{1}{4} \times \frac{1}{3} = \frac{?}{?}$

Note taking

2e

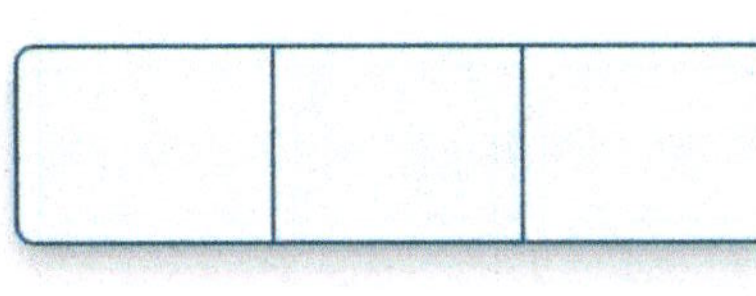

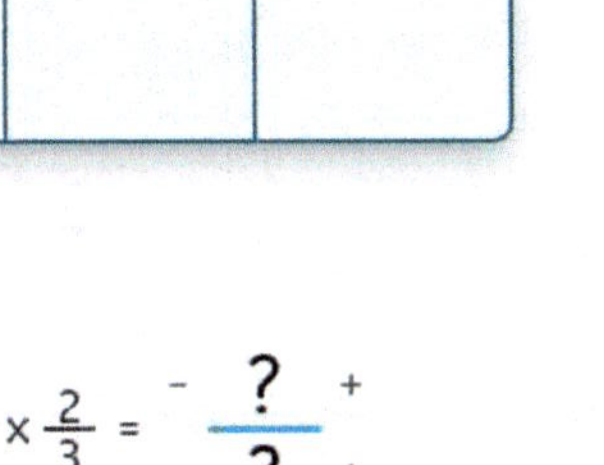

$\frac{1}{2} \times \frac{2}{3} = \frac{?}{?}$

Note taking

2f

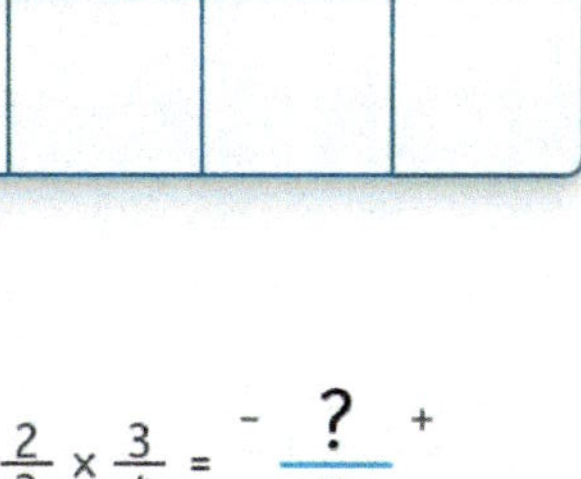

$\frac{2}{3} \times \frac{3}{4} = \frac{?}{?}$

Note taking

2g

$\frac{1}{4} \times \frac{2}{5} = \frac{?}{?}$

Simplify, if you can.

Note taking

2h

$\frac{5}{6} \times \frac{2}{5} = \frac{?}{?}$

Simplify, if you can.

Note taking

2i

$\frac{3}{4} \times \frac{3}{5} = \frac{?}{?}$

Simplify, if you can.

Note taking

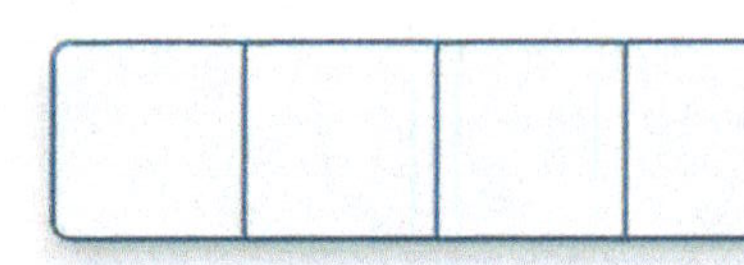

2j

$\frac{4}{7} \times \frac{2}{4} = \frac{?}{?}$

Simplify, if you can.

Note taking

2k

$\frac{2}{5} \times \frac{2}{3} = \frac{?}{?}$

Simplify, if you can.

Note taking

1a

Do You Remember?

$\frac{1}{4} \times \frac{4}{5} = \frac{?}{?}$

Note taking

1b

Do You Remember?

$\frac{1}{2} \times \frac{2}{7} = \frac{?}{?}$

Note taking

1c

Do You Remember?

$\frac{1}{4} \times \frac{4}{9} = \frac{?}{?}$

Note taking

1d

Do You Remember?

Simplify by dividing by
the largest common denominator.

Roadmap

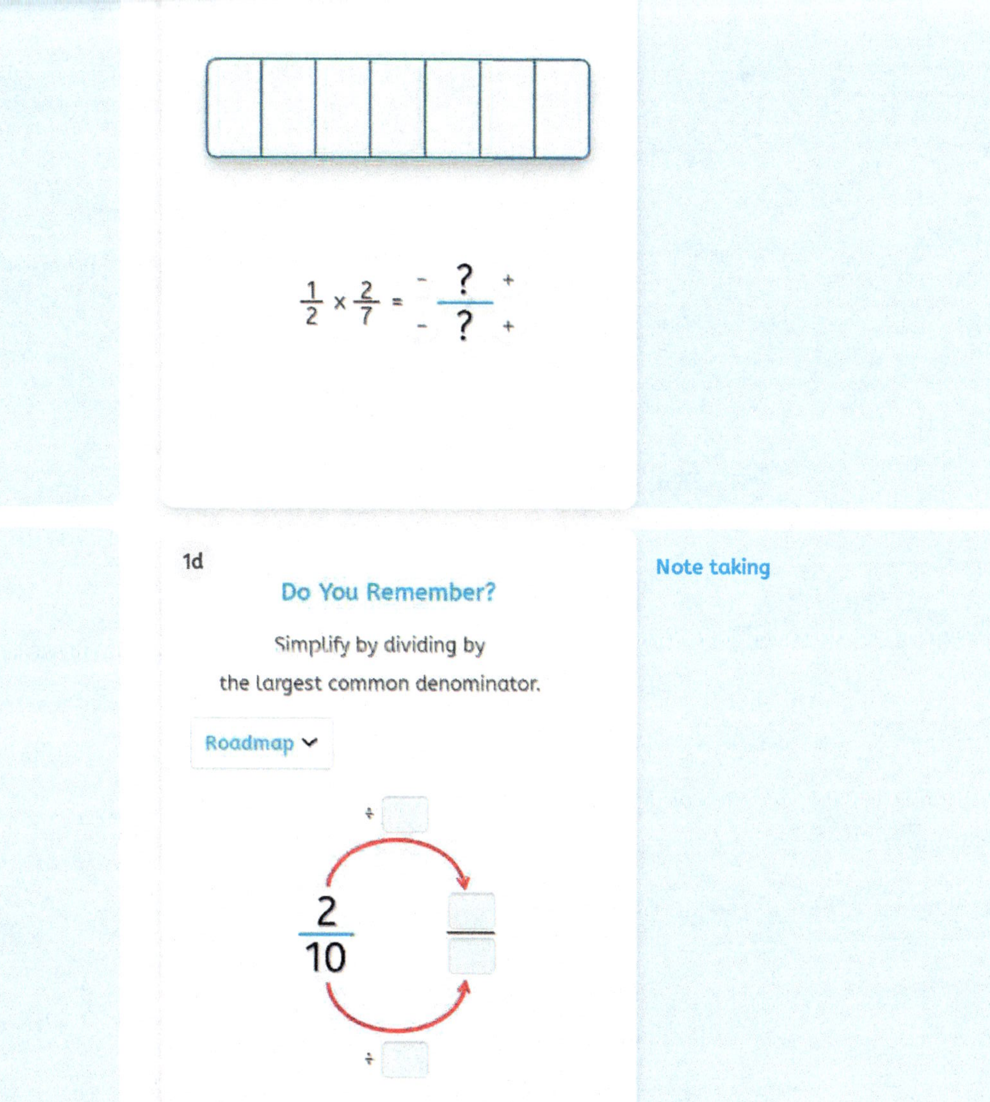

Note taking

1e

Do You Remember?

Simplify by dividing by
the largest common denominator.

Roadmap

$\frac{10}{15}$ ÷ ☐ → $\frac{☐}{☐}$

Note taking

1r Hint

$\frac{1}{2} \times \frac{3}{7} = \ldots$

num. x num. = ☐ x ☐ = ☐
denom. x denom. = ☐ x ☐ = ☐

$\frac{1}{2} \times \frac{3}{7} = \frac{?}{?}$

Note taking

1s Hint

$\frac{2}{5} \times \frac{3}{4} = \ldots$

nom. x nom. = ☐ x ☐ = ☐
denom. x denom. = ☐ x ☐ = ☐

Simplify the answer.

$\frac{2}{5} \times \frac{3}{4} = \frac{?}{?}$

Note taking

1t

Simplify the answer.

$\frac{1}{6} \times \frac{2}{3} = \frac{?}{?}$

Note taking

1u

Simplify the answer.

$\frac{2}{3} \times \frac{5}{8} = \frac{?}{?}$

Note taking

2a

$\frac{1}{2} \times \frac{1}{4} = \frac{?}{?}$

Note taking

2b int

$\frac{1}{2} \times \frac{1}{4} = \ldots$

num. x num. = ☐ x ☐ = ☐
denom. x denom. = ☐ x ☐ = ☐

$\frac{1}{2} \times \frac{1}{4} = \frac{?}{?}$

Note taking

2c int

$\frac{1}{4} \times \frac{1}{3} = \ldots$

num. x num. = ☐ x ☐ = ☐
denom. x denom. = ☐ x ☐ = ☐

$\frac{1}{4} \times \frac{1}{3} = \frac{?}{?}$

Note taking

2d int

$\frac{1}{3} \times \frac{2}{5} = \dots$

num. x num. = $\frac{\square}{\square} \times \frac{\square}{\square} = \frac{\square}{\square}$
denom. x denom. =

$\frac{1}{3} \times \frac{2}{5} = \frac{?}{?}$

Note taking

2e int

$\frac{1}{2} \times \frac{3}{4} = \dots$

num. x num. = $\frac{\square}{\square} \times \frac{\square}{\square} = \frac{\square}{\square}$
denom. x denom. =

$\frac{1}{2} \times \frac{3}{4} = \frac{?}{?}$

Note taking

2f int

$\frac{2}{7} \times \frac{1}{3} = \dots$

num. x num. = $\frac{\square}{\square} \times \frac{\square}{\square} = \frac{\square}{\square}$
denom. x denom. =

$\frac{2}{7} \times \frac{1}{3} = \frac{?}{?}$

Note taking

2g int

$\frac{2}{3} \times \frac{3}{4} = \dots$

num. x num. = $\frac{\square}{\square} \times \frac{\square}{\square} = \frac{\square}{\square}$
denom. x denom. =

$\frac{2}{3} \times \frac{3}{4} = \frac{?}{?}$

Note taking

2h

Simplify the answer.

$\frac{1}{3} \times \frac{2}{8} = \frac{?}{?}$

Note taking

2i

Simplify the answer.

$\frac{2}{5} \times \frac{1}{6} = \frac{?}{?}$

Note taking

2j

Simplify the answer.

$\frac{7}{8} \times \frac{1}{2} = \frac{?}{?}$

Note taking

2k

Simplify the answer.

$\frac{2}{9} \times \frac{3}{4} = \frac{?}{?}$

Note taking

2l

Simplify the answer.

$\frac{3}{4} \times \frac{2}{5} = \frac{?}{?}$

Note taking

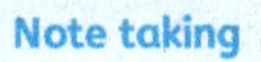

1a

Do you remember?

Write the mixed number as a fraction.

$4\frac{2}{3} = \frac{?}{?}$

Note taking

1b

Do you remember?

Write the mixed number as a fraction.

$3\frac{5}{6} = \frac{?}{?}$

Note taking

1c

Do you remember?

$\frac{1}{6} \times \frac{5}{3} = \frac{?}{?}$

Note taking

1d

Do you remember?

$\frac{2}{3} \times \frac{2}{8} = \frac{?}{?}$

Simplify as necessary.

Note taking

1h

What is $2\frac{1}{2} \times 1\frac{3}{4}$?

Step 1: To find the product when one or more factors are mixed numbers, rewrite each mixed number as a fraction.

$2\frac{1}{2} = \frac{?}{?}$ $1\frac{3}{4} = \frac{?}{?}$

Note taking

1i

What is $2\frac{1}{2} \times 1\frac{3}{4}$?

Step 2: Multiply the fractions.

$\frac{5}{2} \times \frac{7}{4} = \frac{?}{?}$

Note taking

1j

What is $2\frac{1}{2} \times 1\frac{3}{4}$?

Step 3: Write the product as a mixed number.

$\frac{35}{8} = \square\frac{\square}{\square}$

Note taking

1k

Find the product.

Steps

$3\frac{1}{2} \times 2\frac{1}{2} = \frac{\square}{\square} \times \frac{\square}{\square}$

$= \frac{\square}{\square} = \square\frac{\square}{\square}$

Fill in all the boxes.

Note taking

1l Find the product.

Steps

$3\frac{1}{2} \times 1\frac{2}{3} = \frac{\square}{\square} \times \frac{\square}{\square}$

$= \frac{\square}{\square} = \square\frac{\square}{\square}$

Fill in all the boxes.

Note taking

1m Find the product.

Steps

$2\frac{1}{2} \times 1\frac{3}{8} = \square\frac{\square}{\square}$

Note taking

1n Find the product.

$4\frac{1}{2} \times 1\frac{2}{6} = \square$

Note taking

2a Find the product.

Steps

$6\frac{1}{5} \times 1\frac{1}{2} = \frac{\square}{\square} \times \frac{\square}{\square}$

$= \frac{\square}{\square} = \square\frac{\square}{\square}$

Fill in all the boxes.

Note taking

2b Find the product.

Steps

$2\frac{1}{3} \times 1\frac{3}{4} = \frac{\square}{\square} \times \frac{\square}{\square}$

$= \frac{\square}{\square} = \square\frac{\square}{\square}$

Fill in all the boxes.

Note taking

2c Find the product.

Steps

$4\frac{1}{3} \times 1\frac{3}{4} = \square\frac{\square}{\square}$

Note taking

2d Find the product.

Steps

$5\frac{1}{2} \times 2\frac{1}{2} = \square\frac{\square}{\square}$

Note taking

2e Find the product.

Steps

$1\frac{3}{8} \times 1\frac{2}{3} = \square\frac{\square}{\square}$

Note taking

2f

Find the product.

$2\frac{1}{2} \times 2\frac{1}{3} = \square\frac{\square}{\square}$

Note taking

2g

Find the product.

$1\frac{3}{4} \times 2\frac{1}{6} = \square\frac{\square}{\square}$

Note taking

2h

Find the product.

$2\frac{1}{3} \times 4\frac{1}{3} = \square\frac{\square}{\square}$

Note taking

1a

Do You Remember?

$3 \times \frac{1}{5} = \frac{?}{?}$

Note taking

1b

Do You Remember?

$2 \times \frac{1}{3} = \frac{?}{?}$

Note taking

1h

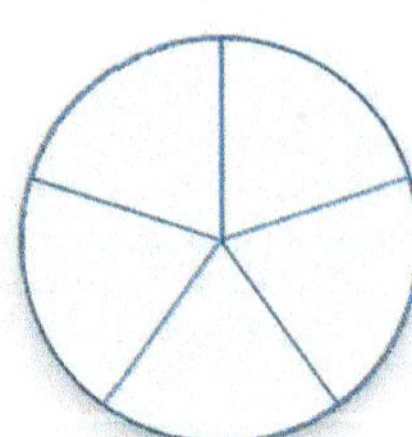

Mira has $\frac{1}{3}$ of a fruit tart.

Of that, she eats $\frac{1}{2}$ part.

$\frac{1}{2}$ part of $\frac{1}{3}$ fruit tart

$\frac{1}{2} \times \frac{1}{3} = \frac{?}{?}$

Note taking

1k

There's still $\frac{1}{4}$ part of a lasagna.

Ryan eats $\frac{1}{2}$ part of that.

What part of the lasagna does Ryan eat?

$\frac{1}{2} \times \frac{1}{4} = \frac{?}{?}$

Note taking

1l

Sam has $\frac{3}{4}$ of a fruit tart.

He eats $\frac{1}{3}$ part of that.

How much does he eat?

$$\frac{1}{3} \times \frac{3}{4} = \frac{?}{?}$$

Note taking

2a

Fernando opened a pizza box. Inside, there was $\frac{3}{4}$ of a pizza.

Fernando ate $\frac{1}{2}$ of the remaining pizza. How much of the pizza did Fernando eat?

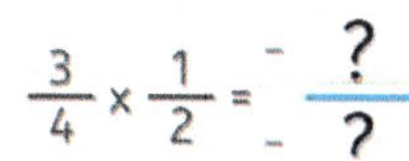

$$\frac{3}{4} \times \frac{1}{2} = \frac{?}{?}$$

Note taking

2b

Fernando opened a pizza box. Inside, there was $\frac{3}{8}$ of a pizza.

Fernando ate $\frac{1}{2}$ of the remaining pizza. How much of the pizza did Fernando eat?

$$\frac{3}{8} \times \frac{1}{2} = \frac{?}{?}$$

Note taking

2c

Maha opened a pizza box. Inside, there was $\frac{5}{8}$ of a pizza.

Maha ate $\frac{1}{4}$ of the remaining pizza. How much of the pizza did Maha eat?

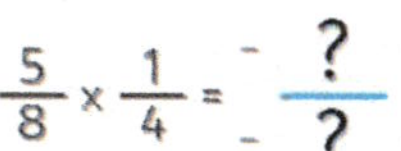

$$\frac{5}{8} \times \frac{1}{4} = \frac{?}{?}$$

Note taking

2d

Dorothy opened a pizza box. Inside, there was $\frac{5}{6}$ of a pizza.

Dorothy ate $\frac{3}{4}$ of the remaining pizza. How much of the pizza did Dorothy eat?

$\frac{5}{6} \times \frac{3}{4} = \frac{?}{?}$

Note taking

2e

Blanche opened a pizza box. Inside, there was $\frac{4}{5}$ of a pizza.

Blanche ate $\frac{2}{3}$ of the remaining pizza. How much of the pizza did Blanche eat?

$\frac{\square}{\square} \times \frac{\square}{\square} = \frac{\square}{\square}$

Note taking

2f

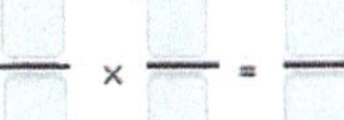

Rose opened a pizza box. Inside, there was $\frac{7}{8}$ of a pizza.

Rose ate $\frac{1}{3}$ of the remaining pizza. How much of the pizza did Rose eat?

$\frac{\square}{\square} \times \frac{\square}{\square} = \frac{\square}{\square}$

Note taking

2g

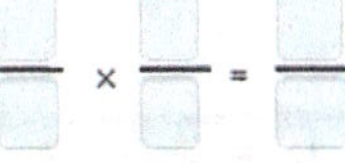

Sofia opened a pizza box. Inside, there was $\frac{1}{4}$ of a pizza.

Sofia ate $\frac{3}{5}$ of the remaining pizza. How much of the pizza did Sofia eat?

$\frac{\square}{\square} \times \frac{\square}{\square} = \frac{\square}{\square}$

Note taking

2h

$\frac{3}{4} \times \frac{3}{5} = \frac{?}{?}$

Brandon opened a bag of carrots. There was $\frac{3}{4}$ of the bag left. Brandon ate $\frac{3}{5}$ of the remaining carrots. How much in carrots did Brandon eat?

Note taking

2i

$\frac{3}{8} \times \frac{2}{4} = \frac{?}{?}$

Faleesa opened a bag of carrots. There was $\frac{3}{8}$ of the bag left. Faleesa ate $\frac{2}{4}$ of the remaining carrots. How much in carrots did Faleesa eat?

Note taking

1a

Do You Remember?

The pizza is divided into four pieces.

Each piece is $\frac{?}{?}$ part.

Note taking

1b

Do You Remember?

What part of the pizza is **left**?

$\frac{?}{?}$

Note taking

1c

Do You Remember?

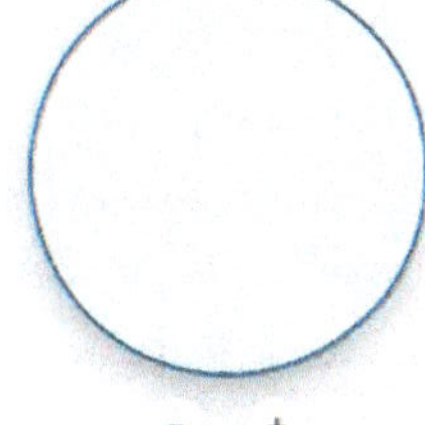

Color $\frac{2}{5}$ part of the circle.

Note taking

1k

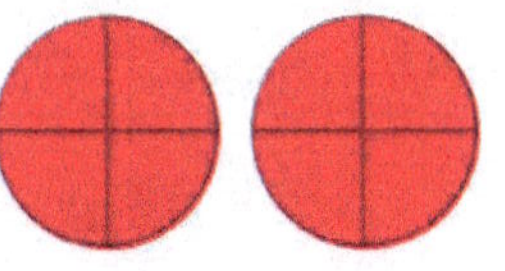

Two pizzas **shared by** four people

Tip

$2 \div 4 = \frac{?}{?}$

Note taking

1l

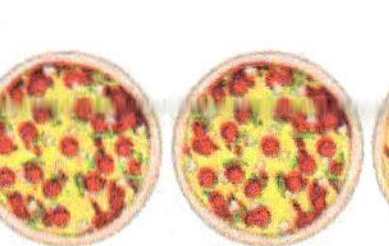
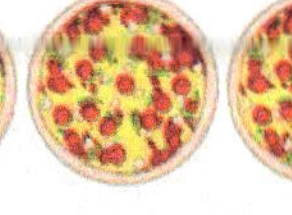

Three pizzas **shared by seven** people

$3 \div 7 = \frac{?}{?}$

Note taking

1m

When do you get $\frac{3}{4}$ pizza?

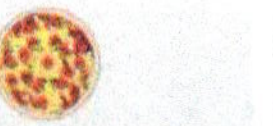

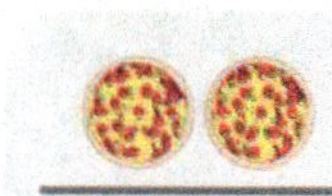

If you share 1 pizza with 4 people.

If you share 4 pizzas with 3 people.

If you share 3 pizzas with 4 people.

If you share 2 pizzas with 3 people.

Note taking

1q

$3 \div 7 = \frac{?}{?}$

Three pizzas shared by seven people

Note taking

1r

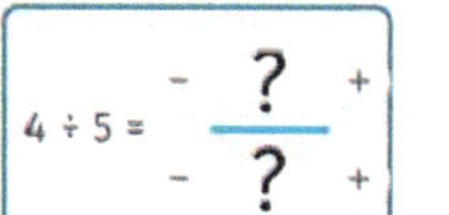

$4 \div 5 = \frac{?}{?}$

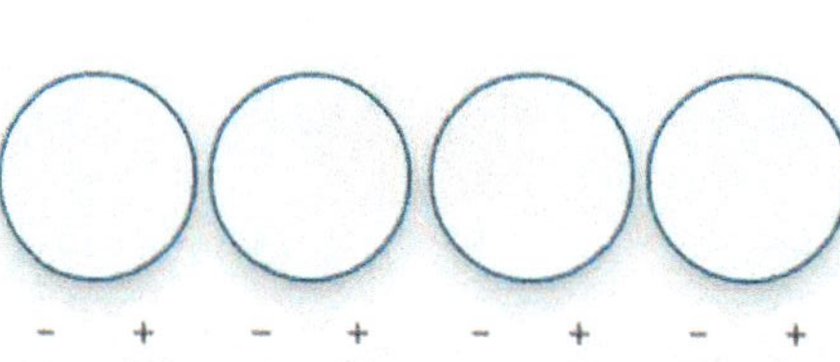

Note taking

1s

$9 \div 10 = \frac{?}{?}$

Note taking

1t

Twelve people share five pancakes.

They each get $\frac{\square}{\square}$ pancakes.

Note taking

1u

When do you get $\frac{3}{8}$ pancake?

If you share ☐ pancakes with ☐ people.

Note taking

2a

Three pizzas **shared by** five people

Tip

$3 \div 5 = \frac{?}{?}$

Note taking

2b

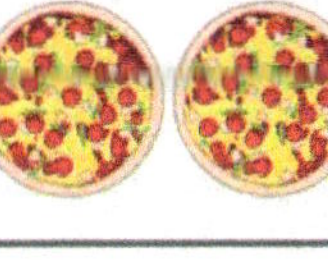

Two pizzas **shared by seven** people

$2 \div 7 = \frac{?}{?}$

Note taking

2c

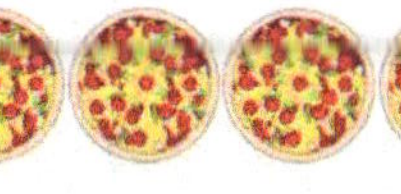

Five pizzas **shared by six** people

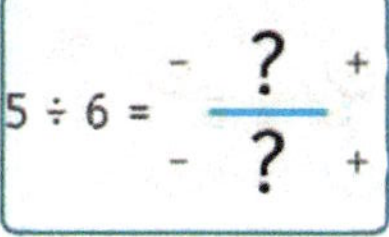

$5 \div 6 = \frac{?}{?}$

Note taking

2d

When do you get $\frac{2}{3}$ pizza?

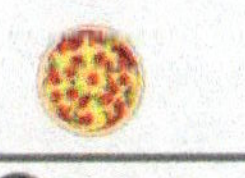

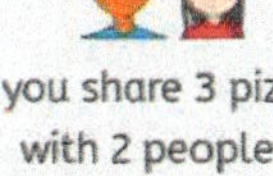

If you share 1 pizza with 3 people.

If you share 3 pizzas with 2 people.

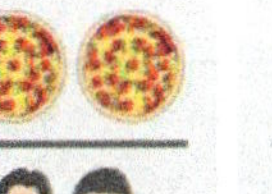

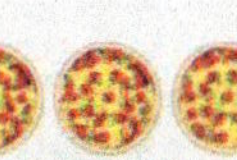

If you share 3 pizzas with 3 people.

If you share 2 pizzas with 3 people.

Note taking

2e

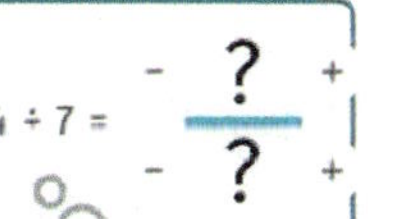

$4 \div 7 = \frac{?}{?}$

Four pizzas shared by seven people

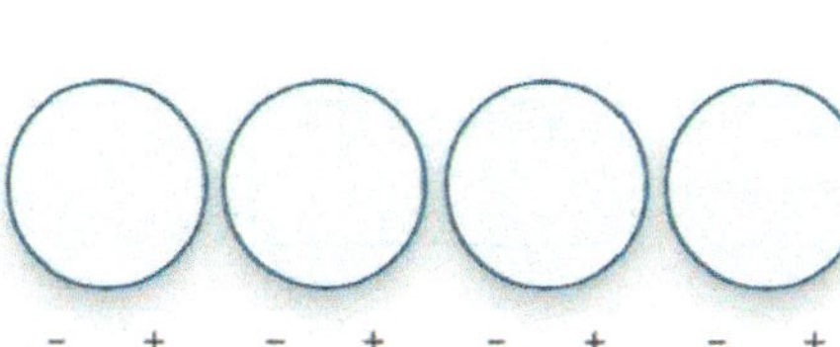

Note taking

2f

$2 \div 9 = \frac{?}{?}$

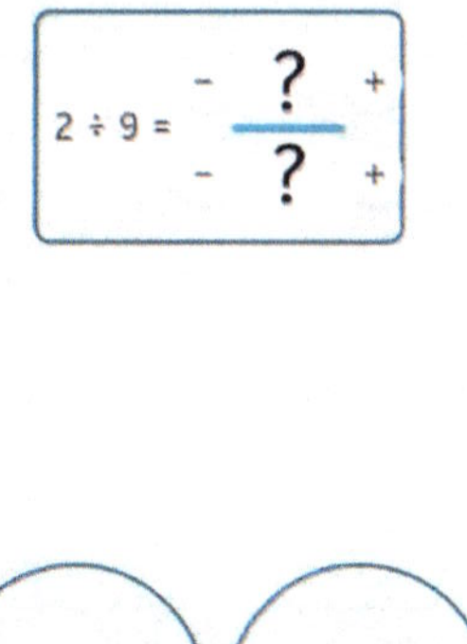

Note taking

2g

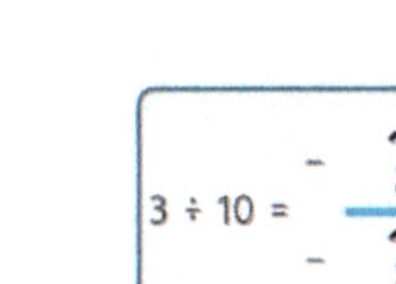

$3 \div 10 = \frac{?}{?}$

Note taking

2h

$6 \div 7 = \frac{?}{?}$

Note taking

2i

Eleven people share four pancakes.

Each person gets $\frac{\square}{\square}$ pancake.

Note taking

2j

When do you get $\frac{4}{9}$ pancake?

If you share ☐ pancakes with ☐ people.

Note taking

1a

Do You Remember?

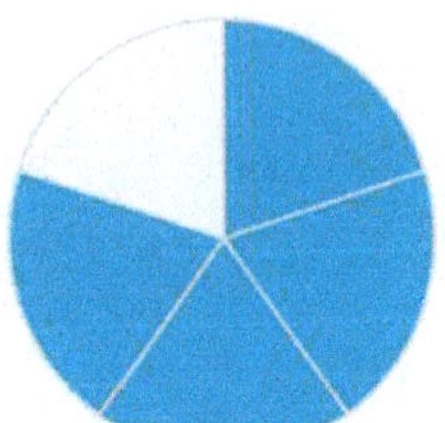

$\frac{1}{4} \times \frac{4}{5} = \frac{?}{?}$

Note taking

1b

Do You Remember?

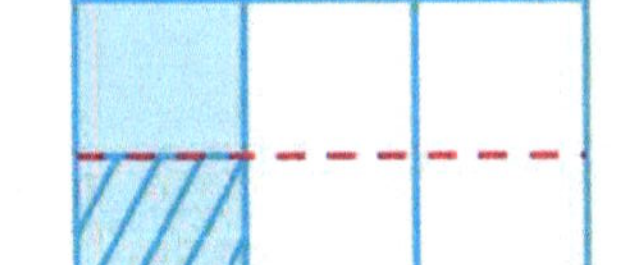
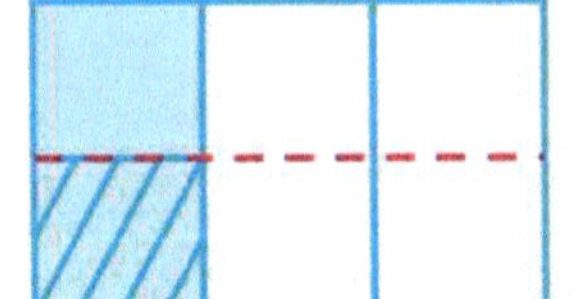

$\frac{1}{2} \times \frac{1}{3} = \frac{?}{?}$

Note taking

1f

Four children share $\frac{4}{5}$ pizza.

How much does each get?

$\frac{4}{5} \div \frac{4}{1} = —$

Each gets — pizza.

Note taking

1i

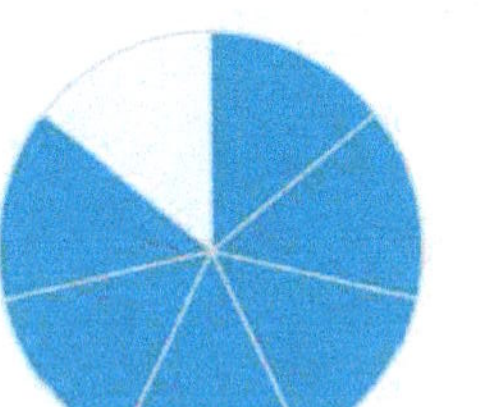

$\frac{6}{7} \div \frac{3}{1} = —$

Note taking

1j

$\frac{5}{6} \div \frac{5}{1} = \frac{}{}$

Note taking

1k

$\frac{8}{9} \div \frac{2}{1} = \frac{}{}$

Note taking

1l

$\frac{3}{5}$

Three children share a pizza.

How much pizza does each child get?

$\frac{3}{5} \div \frac{3}{1} = \frac{?}{?}$

Note taking

1n

$\frac{4}{9} \div \frac{2}{1} = \frac{?}{?}$

Note taking

1o

$\frac{4}{7} \div \frac{4}{1} = \frac{?}{?}$

Note taking

1p

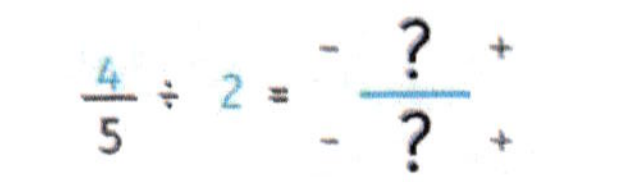

$\frac{4}{5} \div 2 = \frac{?}{?}$

Note taking

2b

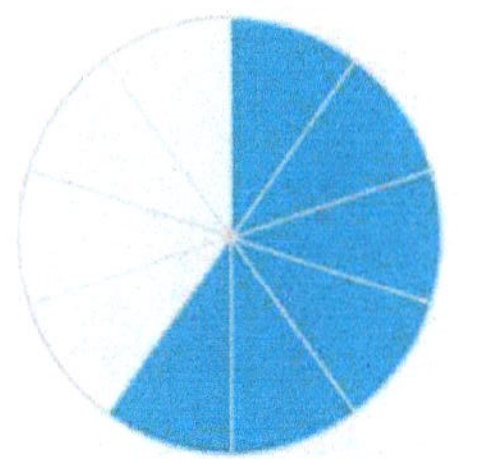

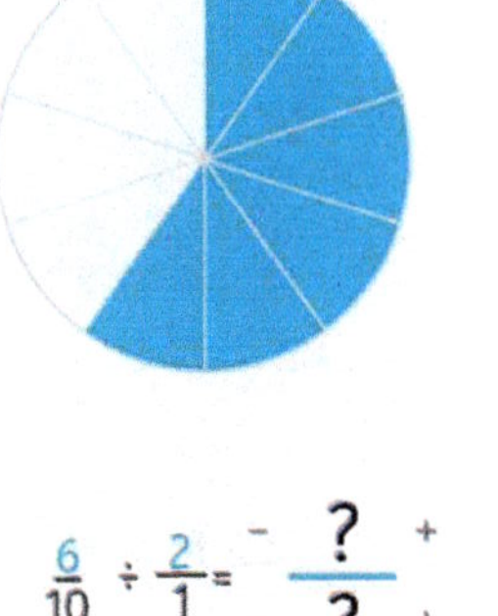

$\frac{6}{10} \div \frac{2}{1} = \frac{?}{?}$

Note taking

2c

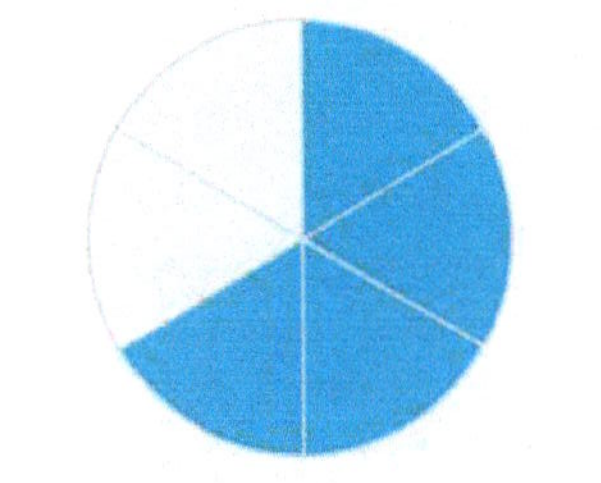

$\frac{4}{6} \div \frac{2}{1} = \frac{?}{?}$

Note taking

2d

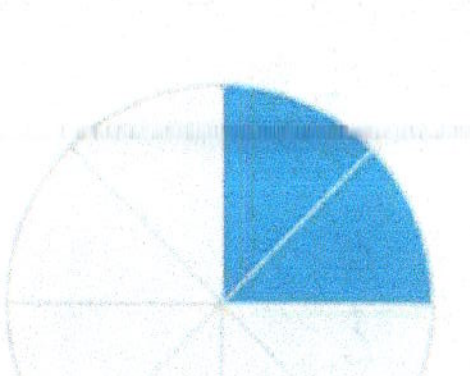

$\frac{2}{8} \div \frac{2}{1} = \frac{?}{?}$

Note taking

2e

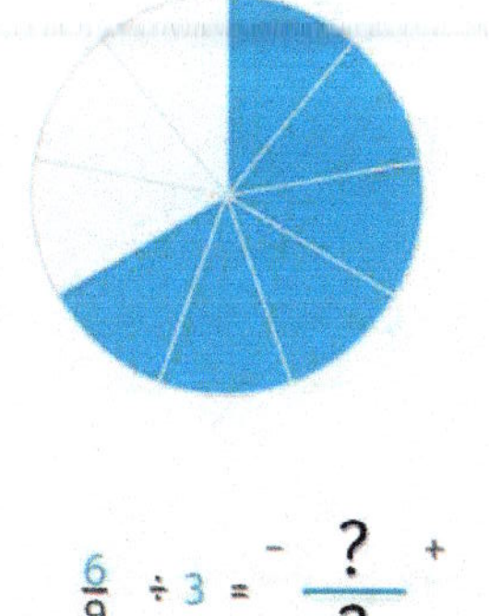

$\frac{6}{9} \div 3 = \frac{?}{?}$

Note taking

2f

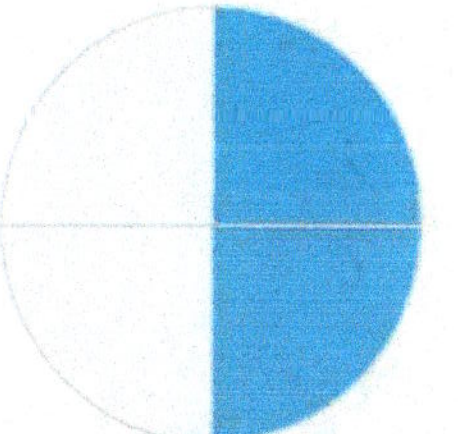

$\frac{2}{4} \div 2 = \frac{?}{?}$

Note taking

2g

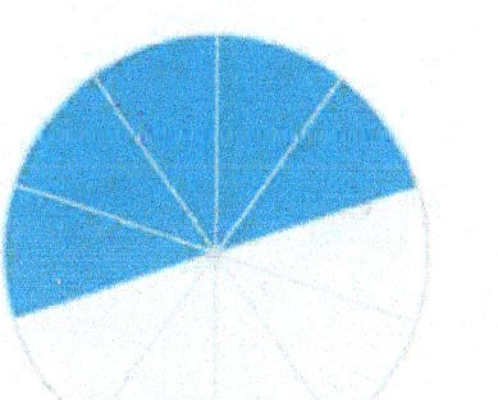

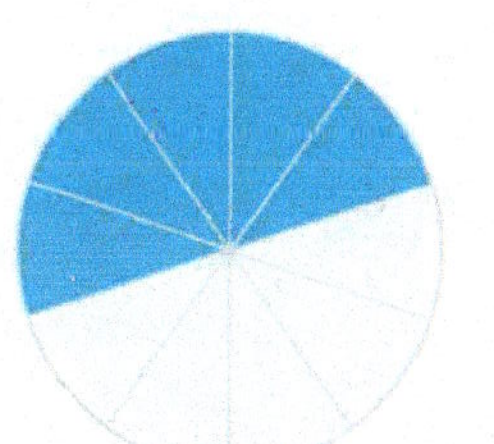

$\frac{2}{10} \div 2 = \frac{?}{?}$

Note taking

2h

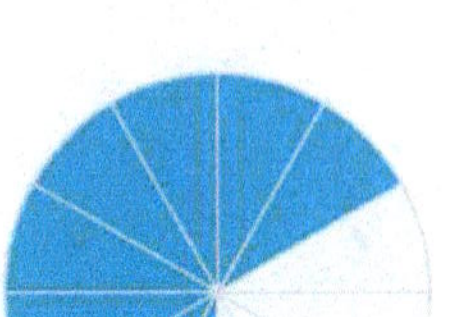

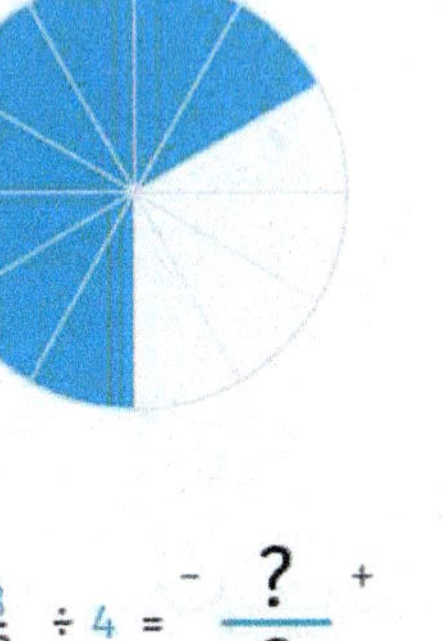

$\frac{8}{12} \div 4 = \frac{?}{?}$

Note taking

2i

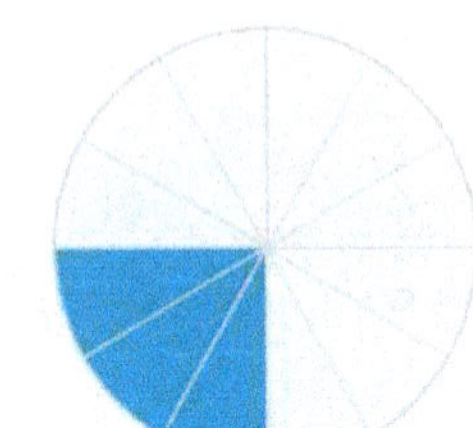

$\frac{3}{12} \div 3 = \frac{?}{?}$

Note taking

2j

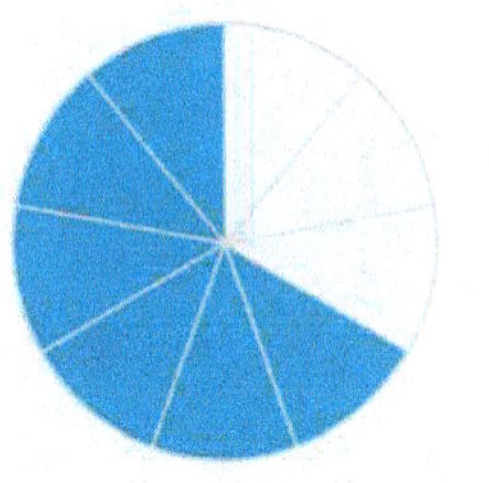

$\frac{6}{9} \div 3 = \frac{?}{?}$

Note taking

2k

$\frac{4}{10} \div 4 = \frac{?}{?}$

Note taking

2l

$\frac{8}{9} \div 8 = \frac{?}{?}$

Note taking

2m

$\frac{8}{12} \div 4 = \frac{?}{?}$

Simplify, if you can.

Note taking

2n

$\frac{10}{16} \div 5 = \frac{?}{?}$

Simplify, if you can.

Note taking

1a

Do You Remember?

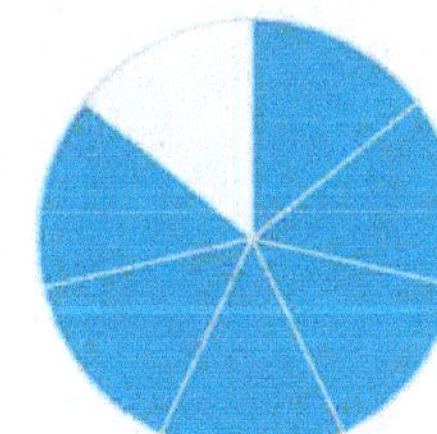

$\frac{6}{7} \div 3 = \frac{?}{?}$

Note taking

1m

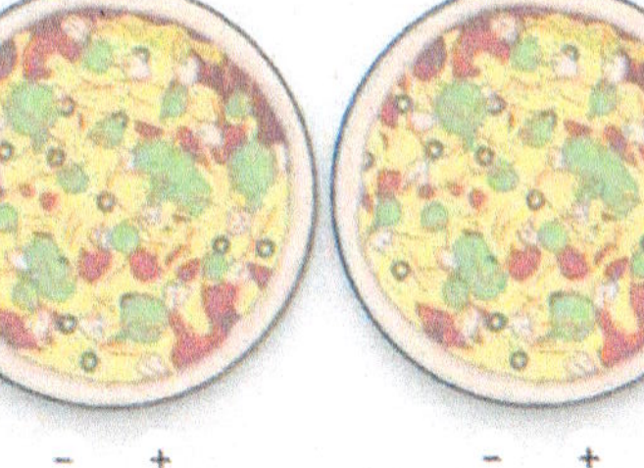

Divide both pizzas into sixths.

How many total pieces of pizza are there?

$2 \div \frac{1}{6} =$ ☐ pieces of pizza

Note taking

1n

$1 \div \frac{1}{3} =$ ☐

$3 \div \frac{1}{3} =$ ☐

Think of a multiplication equation or use the circles below.

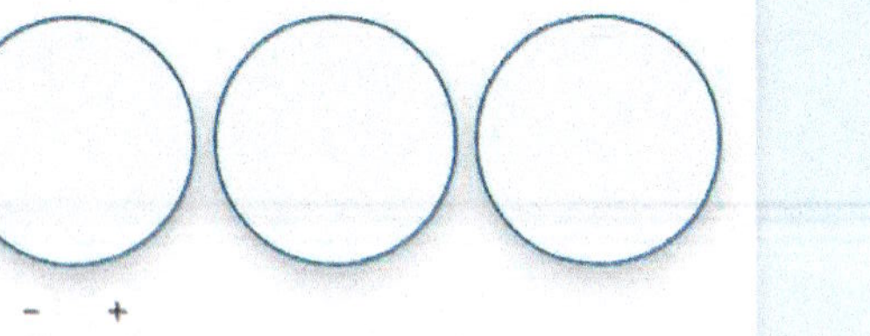

Note taking

1o

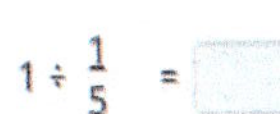

$1 \div \frac{1}{5} =$ ☐

$4 \div \frac{1}{5} =$ ☐

Think of a circle or draw circles on a piece of paper.

Note taking

1p

$5 \div \frac{1}{2} =$ ☐

Think of a circle or draw circles on a piece of paper.

Note taking

1q

$6 \div \frac{1}{4} =$ ☐

Note taking

2a

\- +

Divide the pizza into fifths.

How many pieces of pizza are there?

$1 \div \frac{1}{5} =$ ☐ pieces of pizza

Note taking

2b

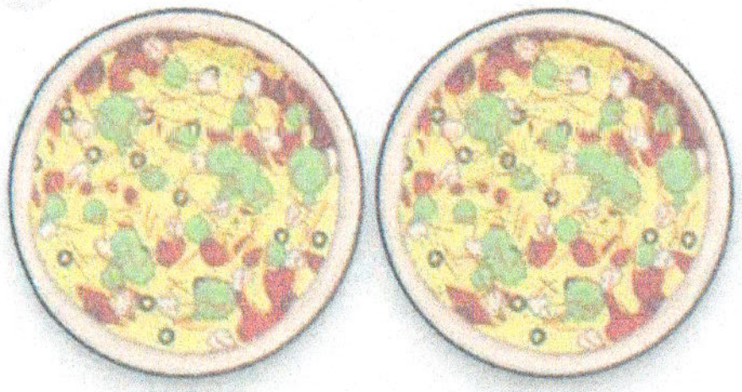

\- + - +

Divide both pizzas into fifths.

How many pieces of pizza are there?

$2 \div \frac{1}{5} =$ ☐ pieces of pizza

Note taking

2c

$1 \div \frac{1}{4} =$ ☐

$2 \div \frac{1}{4} =$ ☐

Think of a circle or use the circles below.

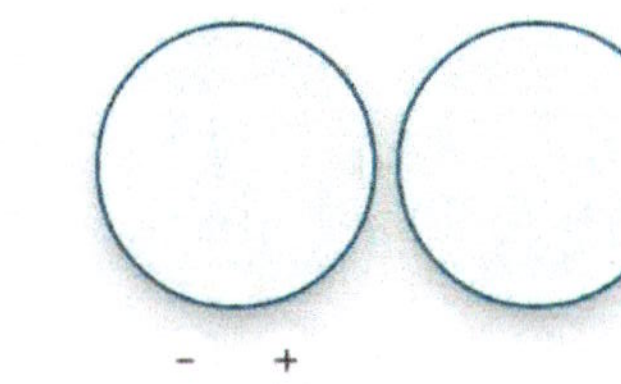
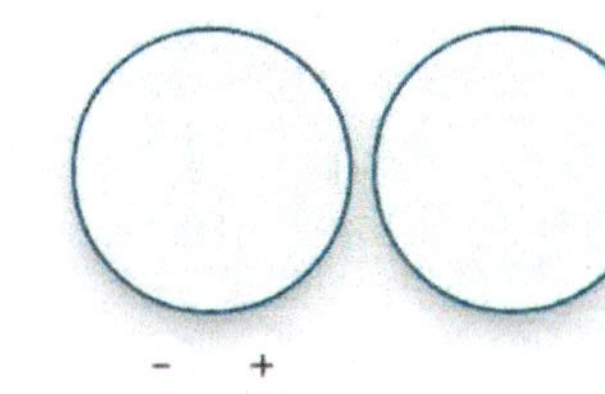

- +

Note taking

2d

$1 \div \frac{1}{6} =$ ☐

$3 \div \frac{1}{6} =$ ☐

Think of a circle or use the circles below.

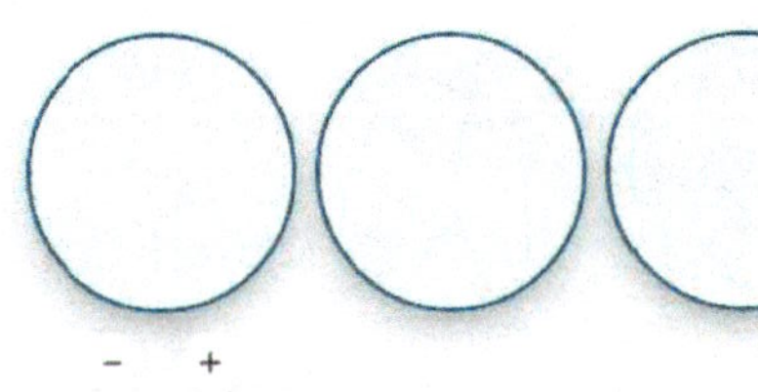

- +

Note taking

2e

$1 \div \frac{1}{3} =$ ☐

$4 \div \frac{1}{3} =$ ☐

Think of a circle or draw circles on a piece of paper.

Note taking

2f

$1 \div \frac{1}{5} =$ ☐

$6 \div \frac{1}{5} =$ ☐

Think of a circle or draw circles on a piece of paper.

Note taking

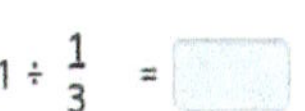

2g

$3 \div \frac{1}{2} =$ ☐

Think of a circle or draw circles on a piece of paper.

Note taking

2h

$6 \div \frac{1}{4} =$ ☐

Think of a circle or draw circles on a piece of paper.

Note taking

2i

$5 \div \frac{1}{3} =$ ☐

Note taking

2j

$2 \div \frac{1}{6} =$ ☐

Note taking

1a

Do you remember?

Jori buys a backpack for $18,
a football for $7, and a
cone for $4.
How much does he pay?

Jori pays $ ☐ .

Note taking

1b

Do you remember?

5 x 7 = ☐

Note taking

1j

Coach Joe bought 5 six-packs and
four single bottles of water for his team.
How many water bottles did he buy in all?

Write an equation: ☐ x 6 + ☐ = ?

Simplify: ☐ + 4 = ?

☐ = ?

Coach Joe bought a total of ☐ water bottles.

Note taking

1k

Jeannie buys

7 meat pizzas and 6 cheese pizzas

12 slices per pizza — 8 slices per pizza

How many slices of pizza does Jeannie buy?
Write an equation and solve.

☐ x 12 + 6 x ☐ =

☐ ☐ 48 = 132

Jeannie buys 132 slices of pizza.

Note taking

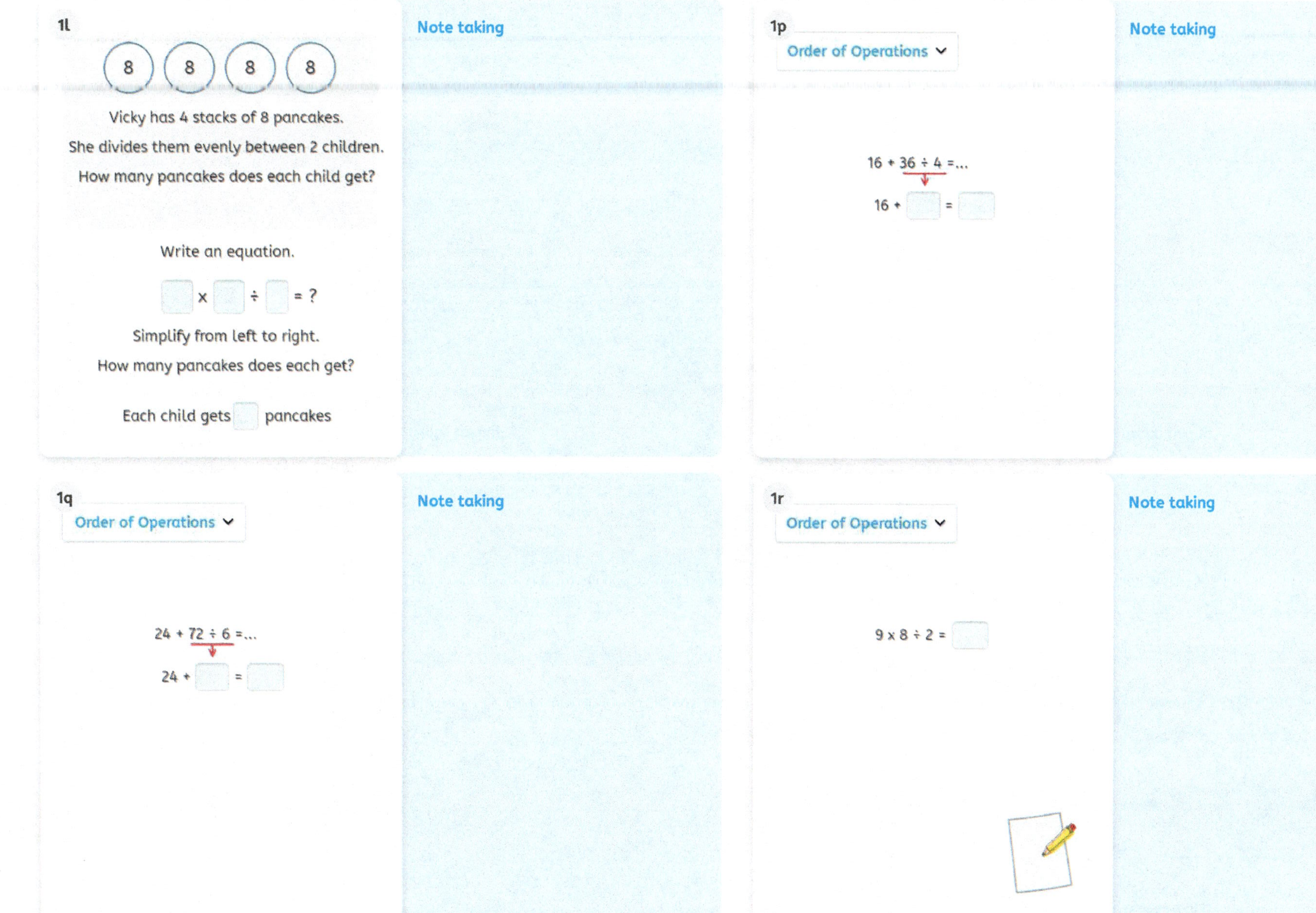

1l

8 8 8 8

Vicky has 4 stacks of 8 pancakes.
She divides them evenly between 2 children.
How many pancakes does each child get?

Write an equation.

☐ × ☐ ÷ ☐ = ?

Simplify from left to right.
How many pancakes does each get?

Each child gets ☐ pancakes

Note taking

1p

Order of Operations

16 + 36 ÷ 4 =...

16 + ☐ = ☐

Note taking

1q

Order of Operations

24 + 72 ÷ 6 =...

24 + ☐ = ☐

Note taking

1r

Order of Operations

9 x 8 ÷ 2 = ☐

Note taking

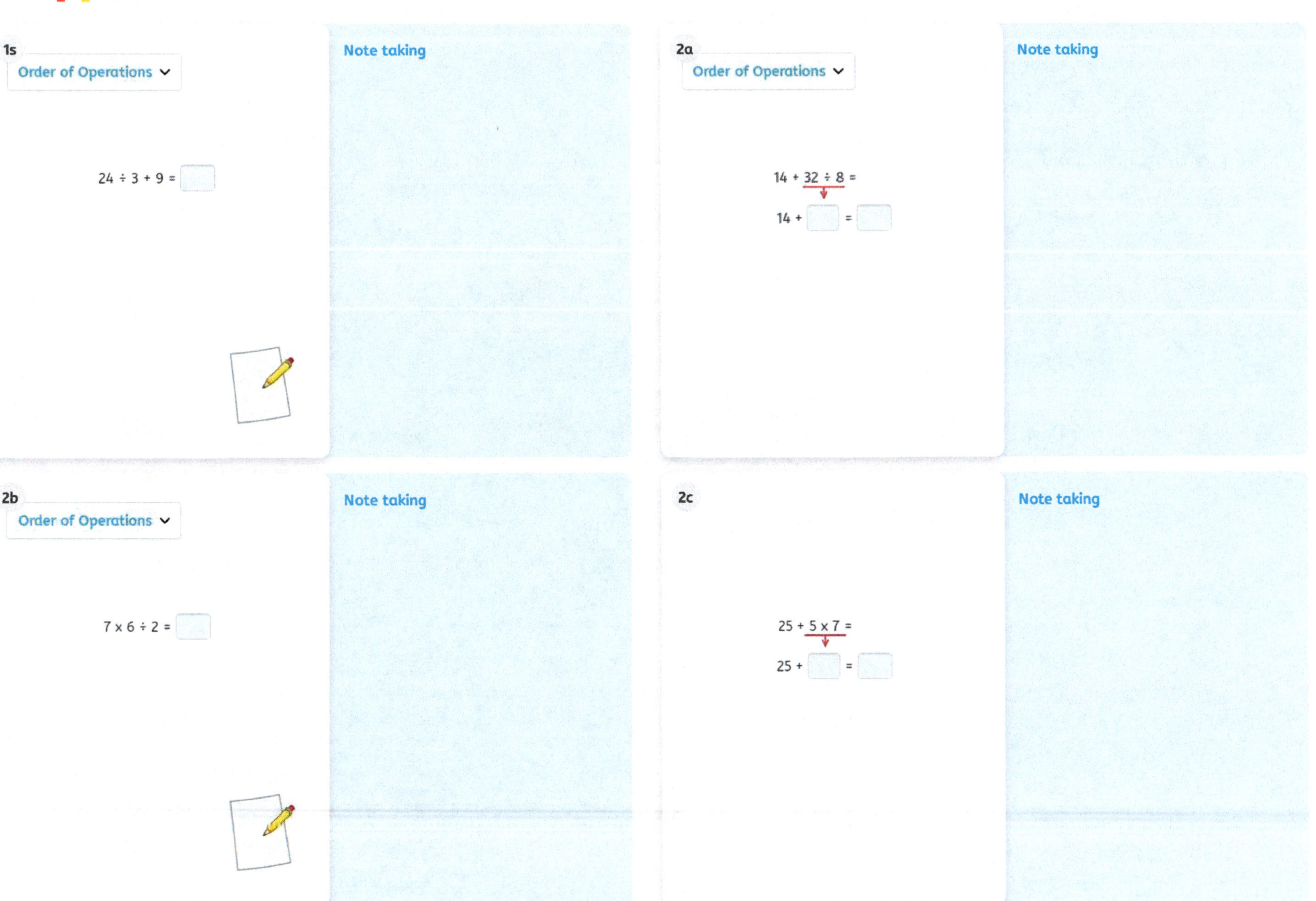

1s

Order of Operations

24 ÷ 3 + 9 = ☐

Note taking

2a

Order of Operations

14 + 32 ÷ 8 =

14 + ☐ = ☐

Note taking

2b

Order of Operations

7 x 6 ÷ 2 = ☐

Note taking

2c

25 + 5 x 7 =

25 + ☐ = ☐

Note taking

2d

60 ÷ 4 - 7 = ☐

Note taking

2e

52 + 9 x 3 = ☐

Note taking

2f

(8 - 6) x (5 + 9) = ...

☐ x ☐ = ☐

Note taking

2g

(54 - 5) ÷ (16 - 9) = ...

☐ ÷ ☐ = ☐

Note taking

2h

$99 - 18 \div 9 =$ ☐

Note taking

2i

$(4 + 7) \times (18 \div 3) =$ ☐

Note taking

2j

$3 \times (63 \div 9) =$ ☐

Note taking

1a Do you remember?

Determine if each statement is true or false.

$10 \div 2 = 2 \div 10$

- true
- false

$10 - 2 = 2 - 10$

- true
- false

$10 \times 2 = 2 \times 10$

- true
- false

$10 + 2 = 2 + 10$

- true
- false

Note taking

1b Do you remember?

$12 \div 2 \times 3 + 5 = ?$

- 7
- 20
- 23
- 48

Note taking

1j Order of Operations

$(9 - 3) \times (7 - 5) =$

☐ × ☐ = ☐

Note taking

1k Order of Operations

$(9 + 3) \div (7 - 1) =$ ☐

Note taking

1l

Order of Operations

$(16 + 36) \div 4 =$ ☐

Note taking

1m

$(7 + 5) \div 6 + 9 =$ ☐

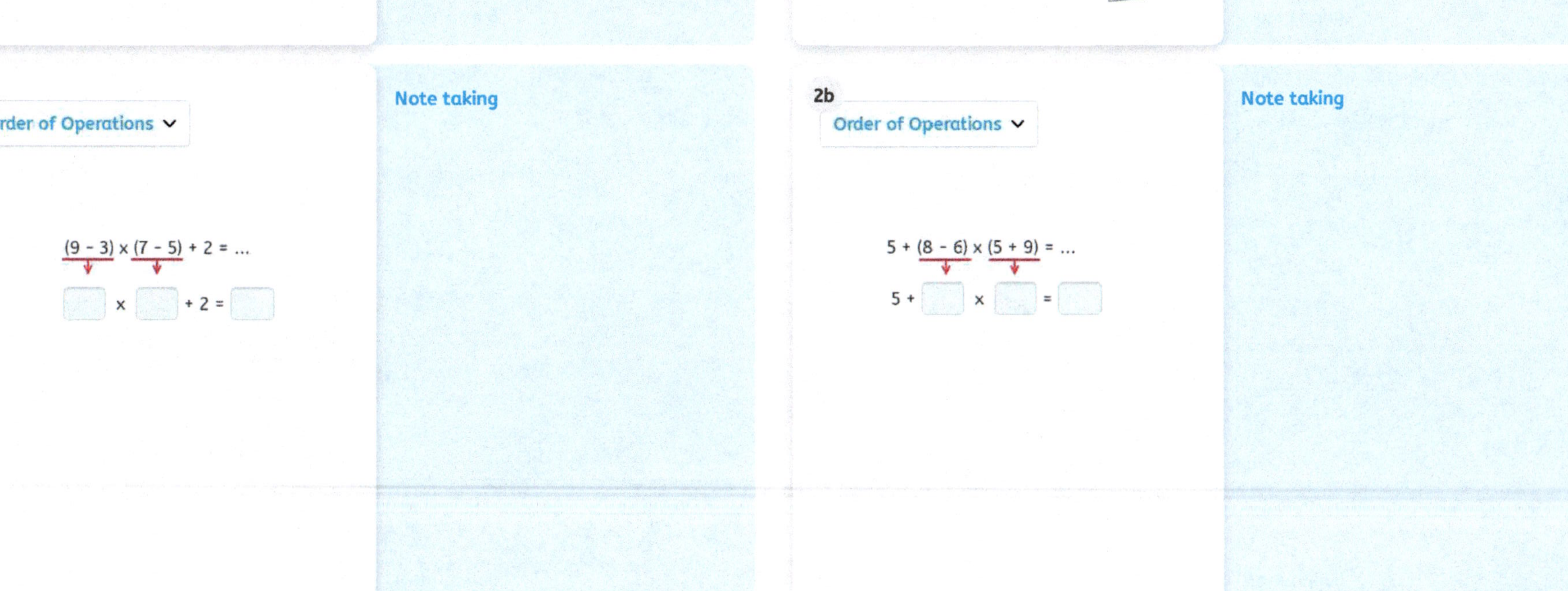

Note taking

2a

Order of Operations

$(9 - 3) \times (7 - 5) + 2 = \ldots$

☐ × ☐ + 2 = ☐

Note taking

2b

Order of Operations

$5 + (8 - 6) \times (5 + 9) = \ldots$

5 + ☐ × ☐ = ☐

Note taking

2c

Order of Operations

$(54 - 6) \div 16 + 9 =$ ☐

Note taking

2d

Order of Operations

$9 + 3 \div (7 - 4) =$ ☐

Note taking

2e

$(7 + 5) \times 6 \div 9 =$ ☐

Note taking

2f

$97 - (18 \div 9) =$ ☐

Note taking

2g

$70 - (4 + 7) \times (18 \div 3) =$ ☐

Note taking

2h

$3 \times (63 \div 9) + 5 =$ ☐

Note taking

2i

Simplify each side to determine if the equation is true.

$10 + 6 \div (3 - 1) = 10 + (6 \div 3) - 1$

☐ ☐

The equation ☐ true.

is

is not

Note taking

2j

Simplify each side to determine if the equation is true.

$(10 \div 5 + 8) \times 6 = (10 \div 5) + 8 \times 6$

☐ ☐

The equation ☐ true.

is

is not

Note taking

1a Do you remember?

$12 \div [(2 \times 3) + 6] = ?$

- 1
- 7
- 21
- 24

Note taking

1b Do you remember?

$4 - \frac{1}{4} = ?$

- 1
- $\frac{3}{4}$
- $3\frac{3}{4}$
- $4\frac{1}{4}$

Note taking

1c Do you remember?

$4 \times \frac{1}{4} = ?$

- 1
- $\frac{3}{4}$
- $3\frac{3}{4}$
- $4\frac{1}{4}$

$4 \div 3 = ?$

- 1
- $\frac{1}{3}$
- $1\frac{1}{3}$
- 12

Note taking

1h Order of Operations

$\frac{1}{2} \times 10 - 4 \div 2 =$

$\frac{\square}{2} - 2 =$

$\square - 2 = \square$

Note taking

1i

$(15 - 3) \div 8 =$

$\square \div 8 = \frac{\square}{8}$

$= \square\,\frac{1}{2}$

Note taking

1j

$(16 \times \frac{1}{4}) \times 3 = \square$

Note taking

2a

Order of Operations

$\frac{1}{2} \times 8 - 12 \div 4 = \square$

Note taking

2b

Order of Operations

$(15 - 3) \div 8 \times 6 =$

$\square \div 8 \times 6 =$

$\frac{\square}{8} \times 6 = \square$

Fill in all the boxes.

Note taking

2c

Order of Operations

$16 \times (\frac{1}{4} + 3) = \square$

Note taking

2d

Order of Operations

$\frac{1}{2} \times 18 - 2 \div 4 = \square \frac{\square}{\square}$

Note taking

2e

Order of Operations

$(16 + 3) \div 4 = \square \frac{\square}{\square}$

Note taking

2f

$3 \times 4 - \frac{1}{4} = ?$

- 3
- $2\frac{1}{4}$
- $11\frac{3}{4}$
- $12\frac{1}{4}$

Note taking

2g

$\frac{1}{3} \times (2 + 7) - 2 = \square$

Note taking

2h

$12 \times \frac{1}{4} \div 4 \times 20 = ?$

- $15\frac{1}{4}$
- $15\frac{3}{4}$
- 15
- 240

Note taking

2i

$20 \times \frac{1}{4} + 3 = ?$

- $\frac{3}{4}$
- 8
- 15
- 65

Note taking

2j

$6 \times \frac{1}{4} + 3 = \square\frac{\square}{\square}$

Note taking

2k

$12 \div [(3 \times 3) + 6] = ?$

- $\frac{3}{5}$
- $\frac{4}{5}$
- $\frac{1}{5}$
- $\frac{2}{5}$

Note taking

1a Do you remember?

Drag and drop a symbol to match each word.

multiply	☐	☐	sum
add	☐	☐	difference
subtract	☐	☐	product
divide	☐	☐	quotient
plus	☐	☐	minus

\+ − × ÷

Note taking

1b Do you remember?

Katy has 6 pencils. She gives some to friends and now has 4 left. How many did she give away?

Write an equation to solve the problem:

6 − ☐ = ☐

Katy gave away ☐ pencils.

Fill in all the boxes.

Note taking

1c Which symbol?

Explanation

"add 4 and 6"

\+ × − ÷

Note taking

1k

"The difference of 9 and 5"

- ◯ 9 − 5
- ◯ 5 − 9
- ◯ 9 + 5
- ◯ 9 × 5

Note taking

1l

"The product of 9 and 5"

- ◯ 9 − 5
- ◯ 5 − 9
- ◯ 9 + 5
- ◯ 9 x 5

Note taking

1m

"The sum of 9 and 5"

- ◯ 9 − 5
- ◯ 5 − 9
- ◯ 9 + 5
- ◯ 9 x 5

Note taking

1n

"The sum of 9 and 5, doubled"

- ◯ (9 − 5) x 2
- ◯ (5 − 9) x 2
- ◯ (9 + 5) x 2
- ◯ (9 x 5) x 2

Note taking

1o

Write an expression for the statement.

"Triple the quotient of 12 and 6"

3 ☐ (☐ ÷ ☐)

Note taking

1p

"The difference of 6 and 1, times the sum of 4 and 2"

(6 ☐ 1) ☐ (4 ☐ 2)

Note taking

1q

"Double the difference of 8 and 5, and add it to 9 divided by 3"

2 ☐ (8 ☐ 5) ☐ (☐ ÷ ☐)

Note taking

1r

"The sum of 10 and 2, cut in half"

- ◯ 10 + 2 ÷ 2
- ◯ 10 + (2 ÷ 2)
- ◯ (10 + 2) ÷ 2
- ◯ (10 + 2) x 2

Note taking

1s

"The quotient of 3 plus 4 and 8 minus 6"

- ◯ 3 + 4 ÷ 8 − 6
- ◯ (3 + 4) ÷ (8 − 6)
- ◯ (3 + 4) ÷ 8 − 6
- ◯ 3 + (4 ÷ 8) − 6

Note taking

2a

"The difference of 8 and 2"

- 8 x 2
- 2 − 8
- 8 + 2
- 8 − 2

Note taking

2b

"The product of 8 and 5"

- 8 − 5
- 5 − 8
- 8 + 5
- 8 x 5

Note taking

2c

"The sum of 9 and 2"

- 9 − 2
- 2 − 9
- 9 + 2
- 9 x 2

Note taking

2d

"The sum of 4 and 9, doubled"

- (4 − 9) x 2
- (9 − 4) x 2
- (4 + 9) x 2
- (4 x 9) x 2

Note taking

2e

Write an expression for the statement.

"Triple the quotient of 16 and 12"

3 ☐ (☐ ÷ ☐)

Note taking

2f

"The difference of 7 and 2, times the sum of 6 and 1"

(7 ☐ 2) ☐ (6 ☐ 1)

Note taking

2g

"Double the sum of 6 and 5, and multiply to 8 divided by 2"

2 ☐ (6 ☐ 5) ☐ (☐ ÷ ☐)

Note taking

2h

"The difference of 9 and 4, cut into 2 equal parts"

- ○ 9 − 4 ÷ 2
- ○ 9 − (4 ÷ 2)
- ○ (9 − 4) ÷ 2
- ○ (9 − 4) x 2

Note taking

2i

"The quotient of 3 plus 4 and 8 minus 6"

- ◯ 3 + 4 ÷ 8 − 6
- ◯ (3 + 4) ÷ (8 − 6)
- ◯ (3 + 4) x 8 − 6
- ◯ 3 + (4 x 8) − 6

Note taking

1a **Do you remember?**

"the difference of 8 and 3, added to the product of 2 and 4"

- ☐ 8 – 3 + 2 x 4
- ☐ (8 – 3) + 2 x 4
- ☐ (8 – 3) + (2 x 4)
- ☐ 3 – 8 + 2 x 4

Note taking

1b **Do you remember?**

"Triple the sum of 6 and 4"

Which operation has to be completed first?

- ○ x
- ○ –
- ○ ÷
- ○ +

Note taking

1c **Which symbol first?**

Explanation

(4 – 2) x (12 ÷ 6) + 1

+ x
– ÷

Note taking

1j Compare the expressions without simplifying.

Expression 1	Expression 2
4 + 5 x 4	(4 + 5) x 4

- ○ Expression 1 is greater than Expression 2.
- ○ Expression 1 is less than Expression 2.

Note taking

1k

What would ***triple*** the value of the given expression?

4 − 1 x 5 + (7 + 2)

- ◯ Multiply 4 by 3
- ◯ Multiply (7 + 2) by 3
- ◯ Multiply the expression by 3
- ◯ Multiply (1 x 5) by 3

Note taking

1l

What would split the given expression into 4 equal parts?

(4 + 3) x (9 − 5) + 1

- ◯ Add 3 to the expression
- ◯ Add 4 to the expression
- ◯ Multiply the expression by 4
- ◯ Divide the expression by 4

Note taking

2a

Compare the expressions without simplifying.

Expression 1	Expression 2
(4 + 5) x 4	4 + 5 x 4

Select ALL that are true.

- ☐ Expression 1 is greater than Expression 2
- ☐ Expression 1 is less than Expression 2
- ☐ The factors for Expression 1 are (4 + 5) and 4
- ☐ The factors for Expression 2 are (4 + 5) and 4

Note taking

2b

Compare the expressions without simplifying.

Expression 1	Expression 2
6 ÷ 3 ÷ 4	6 ÷ 3 ÷ 2

- ◯ Expression 1 is half as large as Expression 2
- ◯ Expression 2 is half as large as Expression 1

Note taking

2c
How can you make the given expression twice as large?

(7 + 3) x 5

- Multiply 7 by 2
- Divide 7 by 2
- Multiply the expression by 2
- Divide the expression by 2

Note taking

2d
What would ***triple*** the value of the given expression?

3 x (2 + 7) − 2

- Multiply 3 by 3
- Multiply (2 + 7) by 3
- Multiply the expression by 3
- Multiply 3 x (2 + 7) by 3

Note taking

2e
What would split the given expression into 3 equal parts?

(4 + 3) x (9 − 5) + 1

- Add 3 to the expression
- Subtract 3 from the expression
- Multiply the expression by 3
- Divide the expression by 3

Note taking

2f
Compare the expressions without simplifying.

Expression 1	Expression 2
(4 + 6) ÷ 3	4 + 6 ÷ 3

- Expression 1 is greater than Expression 2
- Expression 2 is greater than Expression 1

Note taking

2g Compare the expressions without simplifying.

Expression 1	Expression 2
$(4 + 5) \div (3 \times 4) + 3$	$(4 + 5) \div (3 \times 4)$

- The expressions are equivalent
- Expression 1 is greater than Expression 2
- Expression 2 is greater than Expression 1

Note taking

2h What would split the given expression into 3 equal parts?

$$(4 + 3) \times (9 - 5) + 1$$

- Divide the expression by 3
- Multiply the expression by 3
- Divide (4 + 3) by 3
- Multiply (4 + 3) by 3

Note taking

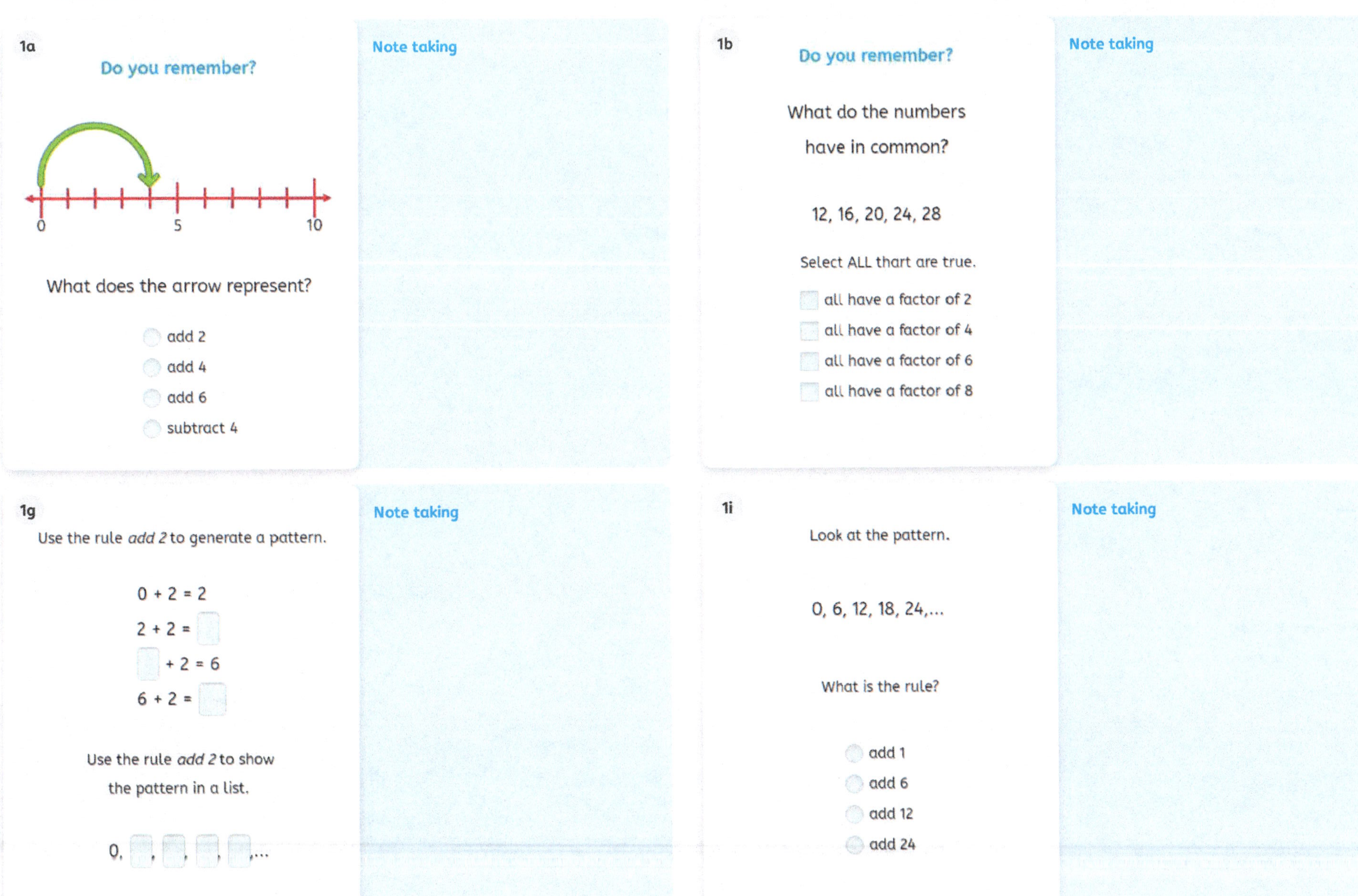

1a

Do you remember?

What does the arrow represent?

- add 2
- add 4
- add 6
- subtract 4

Note taking

1b

Do you remember?

What do the numbers have in common?

12, 16, 20, 24, 28

Select ALL thart are true.

- all have a factor of 2
- all have a factor of 4
- all have a factor of 6
- all have a factor of 8

Note taking

1g

Use the rule *add 2* to generate a pattern.

0 + 2 = 2
2 + 2 = ☐
☐ + 2 = 6
6 + 2 = ☐

Use the rule *add 2* to show the pattern in a list.

0, ☐, ☐, ☐, ☐,...

Note taking

1i

Look at the pattern.

0, 6, 12, 18, 24,...

What is the rule?

- add 1
- add 6
- add 12
- add 24

Note taking

1j

Look at the pattern.

0, 11, 22, 33, 44,...

What is the rule?

- add 1
- add 10
- add 11
- add 22

Note taking

1k

Use the rule to complete the pattern.

add 4

0, ☐, ☐, ☐, ☐, ☐,...

Note taking

1l

Use the rule to complete the pattern.

add 3

0, ☐, ☐, ☐, ☐, ☐,...

Note taking

1m

Look at the pattern.

What is the rule?

0, 10, 20, 30, 40, 50,...

The rule is *add* ☐.

Note taking

2a

Look at the pattern.

0, 12, 24, 36, 48,...

What is the rule?

- ◯ add 2
- ◯ add 4
- ◯ add 10
- ◯ add 12

Note taking

2b

Look at the pattern.

0, 3, 6, 9, 12,...

What is the rule?

- ◯ add 3
- ◯ add 6
- ◯ add 9
- ◯ add 12

Note taking

2c

Look at the pattern.

What is the rule?

0, 1, 2, 3, 4, 5,...

The rule is *add* ☐.

Note taking

2d

Look at the pattern.

What is the rule?

0, 4, 8, 12, 16, 20,...

The rule is *add* ☐.

Note taking

2e

Use the rule to complete the pattern.

add 8

0, ☐, ☐, ☐, ☐, ☐,...

Note taking

2f

Use the rule to complete the pattern.

add 10

0, ☐, ☐, ☐, ☐, ☐,...

Note taking

2g

Analyze and complete the pattern.

0, 2, ☐, ☐, 8, ☐,...

The pattern is *add* ☐.

Note taking

2h

Analyze and complete the pattern.

0, ☐, ☐, 9, ☐, 15,...

The pattern is *add* ☐.

Note taking

2i

Analyze and complete the pattern.

0, 5, ☐, 15, ☐, ☐,...

The pattern is *add* ☐.

Note taking

1a

Do you remember?

0 5 10

What does each arrow represent?

- ◯ add 2
- ◯ add 4
- ◯ add 6
- ◯ subtract 4

Note taking

1b

Do you remember?

What do the numbers have in common?

12, 16, 20, 24, 28

Select ALL that are true.

- ☐ all have a factor of 2
- ☐ all have a factor of 4
- ☐ all have a factor of 6
- ☐ all have a factor of 8

Note taking

1e

Write a rule for the green arrow.

Rule: ☐ 2

Generate more numbers using the rule:

0, ☐, 4, ☐, ☐, 10, ...

Note taking

1f

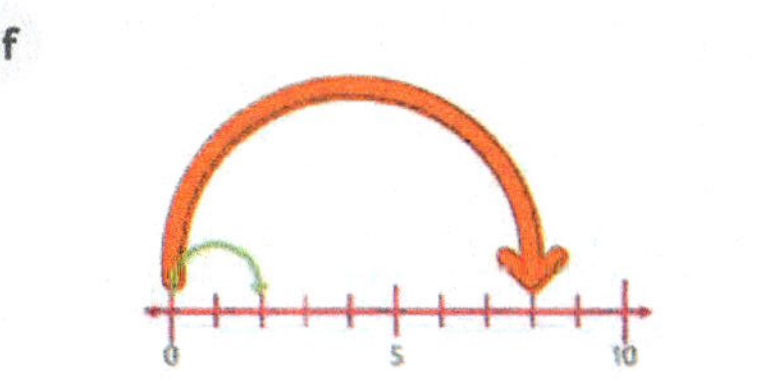

Write a rule for the orange arrow.

Rule: ☐ 8

Use the rule to generate terms:

0, ☐, 16, ☐, 32, ☐, ☐, ...

Note taking

1i

Write a rule for the green arrow.

Rule: *Add* ☐

Describe how you can use this rule to generate more terms.

Note taking

1j

Write a rule for the orange arrow.

Rule: *Add* ☐

How does the rule help you find more numbers?

Note taking

1k **Compare the rules *add 2* and *add 6*.**

Step 1: Make a pattern using the rule *add 2* starting with 0. Then, make a pattern using the rule *add 6* starting with 0.

Add 2	0	2	☐	☐	☐
Add 6	0	6	☐	☐	☐

Step 2: Compare the patterns.

Each number is ☐ times the number above.

The number 6 is ☐ times the number 2.

The rule *add 6* is ☐ times the rule *add 2*.

Note taking

1l **Compare the rules *add 5* and *add 10*.**

Steps

Add 5	0	5	☐	☐	☐
Add 10	0	10	☐	☐	☐

Each number is one-half the number below.

The number 5 is one-☐ the number 10.

The rule *add 5* is one-☐ the rule *add 10*.

Note taking

1m Compare the rules *add 3* and *add 9.*

Steps

Add 3	0				
Add 9	0				

The rule *add 9* is ___ times the rule *add 3.*

Note taking

2a Compare the rules *add 3* and *add 12.*

Steps

Add 3	0	3			
Add 12	0	12			

Each number is ___ times the number above.

The number 12 is ___ times the number 3.

The rule *add 12* is ___ times the rule *add 3.*

Note taking

2b Compare the rules *add 10* and *add 20.*

Steps

Add 10	0				
Add 20	0				

Each number is one-half the number below.

The number 10 is one-___ the number 20.

The rule *add 10* is one-___ the rule *add 20.*

Note taking

2c Compare the rules *add 2* and *add 10.*

Add 2	0	2	4	6	8
Add 10	0	10	20	30	40

The rule *add 10* is...

- one-half the rule *add 2*
- 2 times the rule *add 2*
- 5 times the rule *add 2*
- 5 more than the rule *add 2*

Note taking

2d **Compare the rules *add 5* and *add 20*.**

Add 5	0	5	10	15	20
Add 20	0	20	40	60	80

The rule *add 20* is...

- one-half the rule *add 5*
- 4 times the rule *add 5*
- 2 times the rule *add 5*
- 15 more than the rule *add 5*

Note taking

2e **Compare the rules *add 6* and *add 12*.**

Add 6	0	6	12	18	24
Add 12	0	12	24	36	48

The rule *add 6* is...

- one-half the rule *add 12*
- 2 times the rule *add 12*
- 2 times the rule *add 6*
- 2 more than the rule *add 12*

Note taking

2f **Compare the rules *add 5* and *add 15*.**

Add 5	0				
Add 15	0				

The rule *add 5* is one-______ the rule *add 15*.

- half
- third
- fourth
- fifth

Note taking

2g **Compare the rules *add 6* and *add 12*.**

Add 6	0				
Add 12	0				

The rule *add 6* is one-______ the rule *add 12*.

- half
- third
- fourth
- fifth

Note taking

2h Compare the rules *add 3* and *add 15*.

Add 3	0				
Add 15	0				

The rule *add 15* is ___ times the rule *add 3*.

- 3
- 5
- 10
- 12

Note taking

2i Compare the rules *add 6* and *add 18*.

The rule *add 6* is ________ the rule *add 18*.

- one-third
- one-sixth
- 3 times
- 6 times

Note taking

2j Compare the rules *add 4* and *add 8*.

The rule *add 8* is ________ the rule *add 4*.

- one-half
- one-fourth
- 2 times
- 4 times

Note taking

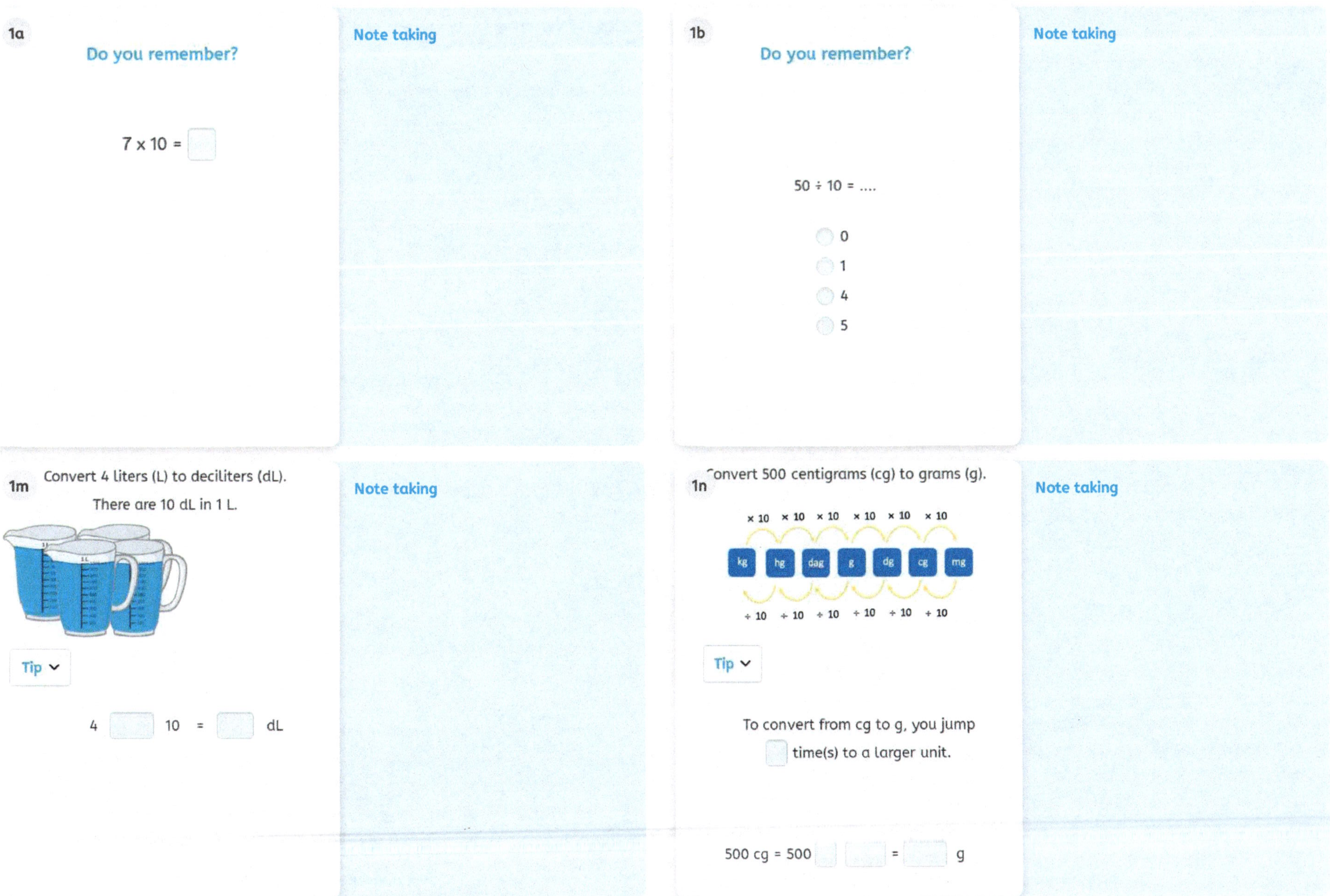

1a Do you remember?

$7 \times 10 =$ ☐

Note taking

1b Do you remember?

$50 \div 10 =$ ....

- 0
- 1
- 4
- 5

Note taking

1m Convert 4 liters (L) to deciliters (dL).

There are 10 dL in 1 L.

Tip

4 ☐ 10 = ☐ dL

Note taking

1n Convert 500 centigrams (cg) to grams (g).

Tip

To convert from cg to g, you jump ☐ time(s) to a larger unit.

500 cg = 500 ☐ ☐ = ☐ g

Note taking

1o vert 150 milligrams (mg) to centigrams (cg).

× 10 × 10 × 10 × 10 × 10 × 10

kg hg dag g dg cg mg

÷ 10 ÷ 10 ÷ 10 ÷ 10 ÷ 10 ÷ 10

mass in mg		mg in 1 cg		mass in cg
150	☐	☐	=	☐

Note taking

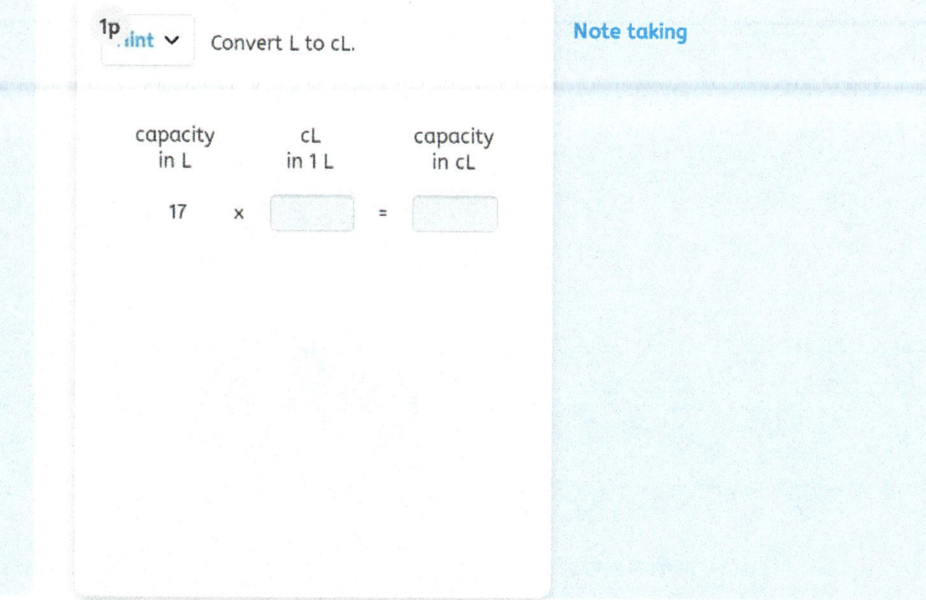

1p Hint Convert L to cL.

capacity in L		cL in 1 L		capacity in cL
17	×	☐	=	☐

Note taking

1r There are 100 centigrams (cg) in 1 gram (g).

Use the table to list the conversions from cg to g.

100 cg	1 g
500 cg	☐ g
800 cg	☐ g
1000 cg	☐ g
1200 cg	☐ g

Note taking

2a Convert 6 kilometers (km) to meters (m).

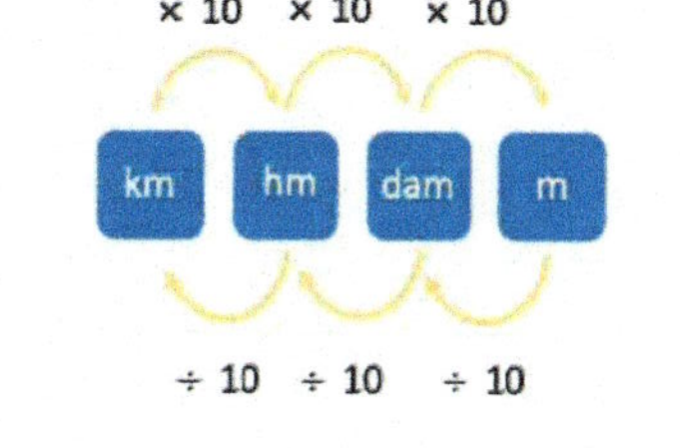

6 km = 6 × ☐ = ☐ m

Note taking

2b

Convert 400 L to daL.

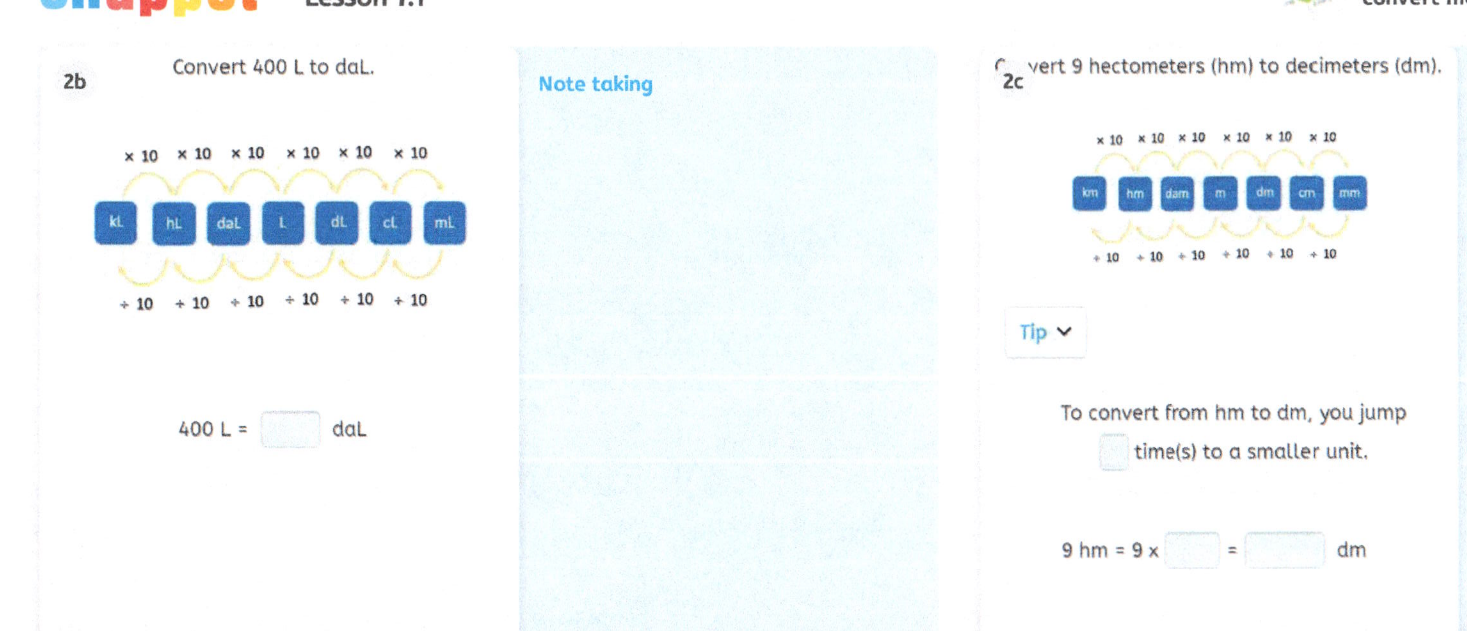

400 L = ☐ daL

Note taking

2c

Convert 9 hectometers (hm) to decimeters (dm).

× 10 × 10 × 10 × 10 × 10 × 10
km hm dam m dm cm mm
÷ 10 ÷ 10 ÷ 10 ÷ 10 ÷ 10 ÷ 10

Tip

To convert from hm to dm, you jump ☐ time(s) to a smaller unit.

9 hm = 9 x ☐ = ☐ dm

Note taking

2d

Hint

5000 g

=

- 500 kg
- 50 kg
- 5 kg

Note taking

2e

Convert 2700 milliliters to deciliters.

There are 100 milliliters in a deciliter.

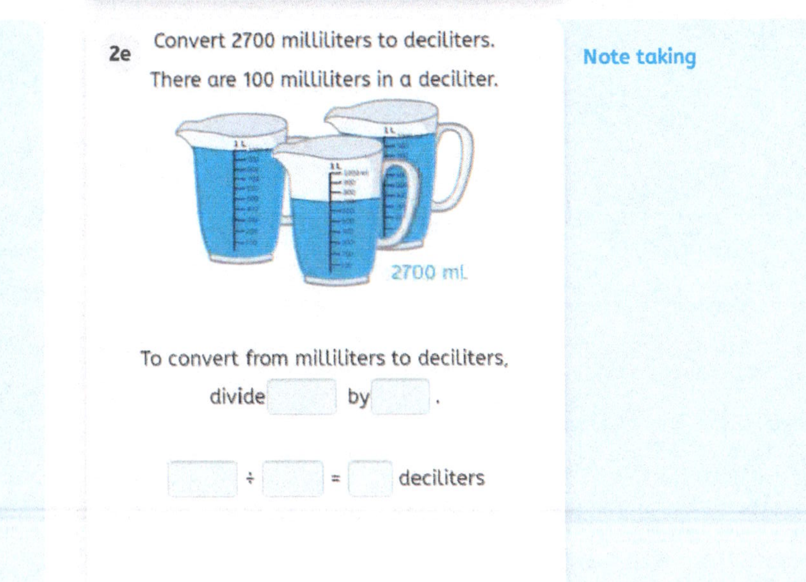

To convert from milliliters to deciliters, divide ☐ by ☐.

☐ ÷ ☐ = ☐ deciliters

Note taking

2f Convert 3500 liters (L) to decaliters (daL).

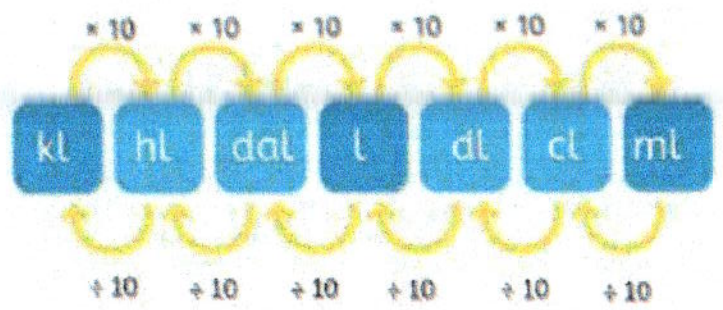

kl hl dal l dl cl ml
× 10 × 10 × 10 × 10 × 10 × 10
÷ 10 ÷ 10 ÷ 10 ÷ 10 ÷ 10 ÷ 10

Tip

3500 L = ☐ daL

Note taking

2g Hint

Convert g to hg.

weight in g		g in 1 hg		weight in hg
320	☐	☐	=	☐

Note taking

2h Hint

Convert L to cL.

capacity in L		cL in 1 L		capacity in cL
500	☐	☐	=	☐

Note taking

2i Hint

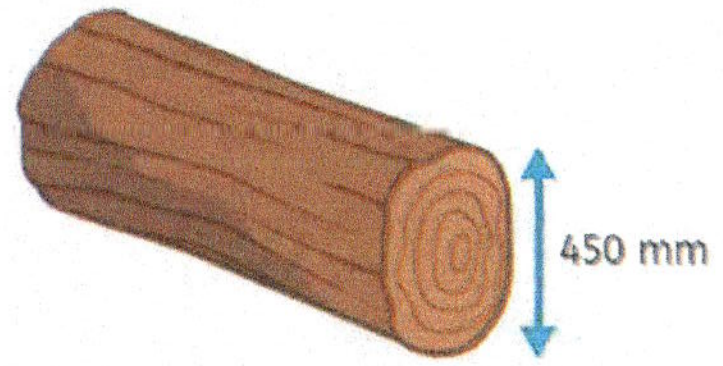

450 mm

=

- 4,500 cm
- 450 cm
- 45 cm

Note taking

2j ·int

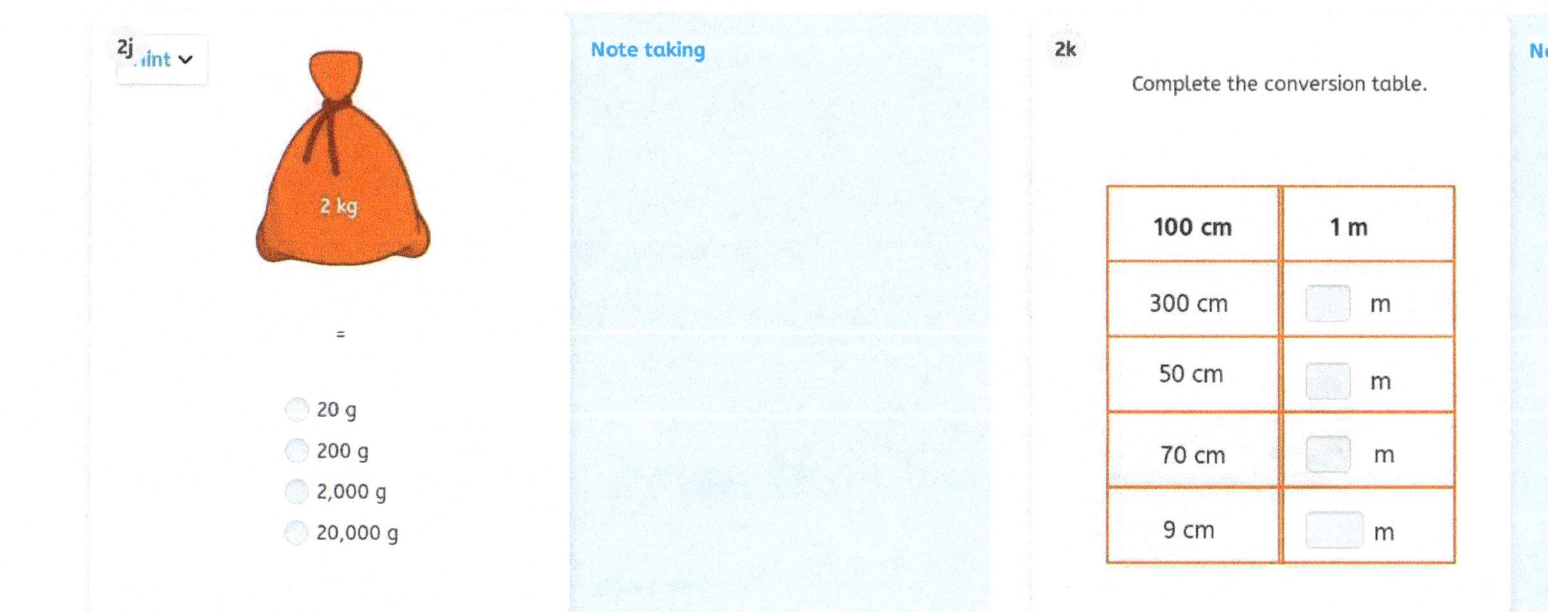

=

- 20 g
- 200 g
- 2,000 g
- 20,000 g

Note taking

2k

Complete the conversion table.

100 cm	1 m
300 cm	m
50 cm	m
70 cm	m
9 cm	m

Note taking

2l

Complete the conversion table.

1000 dg	1 hg
5000 dg	hg
8000 dg	hg
9000 dg	hg
10,000 dg	hg

Note taking

1a **Do you remember?**

Each section of the ruler is 1 inch long.

How long is this rectangle?

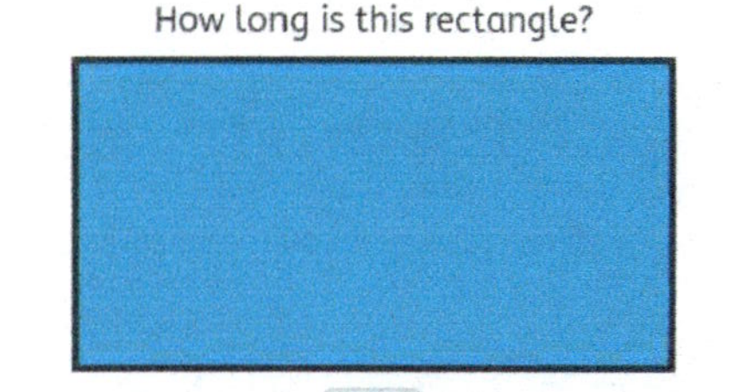

It is about ☐ inches.

drag

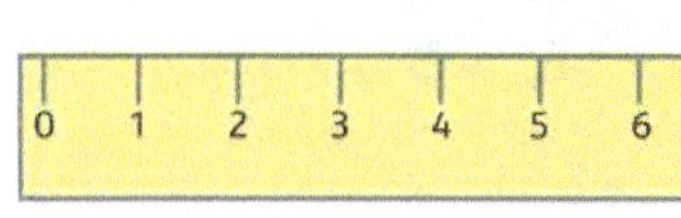

Note taking

1b **Do you remember?**

1 pound (lb) = 16 ounces (oz)

What is the weight in oz?

- 4 oz
- 12 oz
- 20 oz
- 64 oz

Note taking

1i Look at the units in the picture.

How many ounces are there in 5 pounds?

There are ☐ times 16 ounces.

☐ x 16 = ☐ ounces

Note taking

1m

Complete the conversion table.

1 ft	12 in.
2 ft	☐ in.
3 ft	☐ in.
4 ft	☐ in.
5 ft	☐ in.

Note taking

1n Complete the conversion table.

2,000 lb	**1 t**
4,000 lb	☐ t
6,000 lb	☐ t
10,000 lb	☐ t
20,000 lb	☐ t

Note taking

1o Complete the conversion table.

1 gal	☐ qt
2 gal	**8 qt**
3 gal	☐ qt
5 gal	☐ qt
8 gal	☐ qt

Note taking

2a Look at the units in the picture.
How many ounces are there in 9 pounds?

1 lb
16 oz

There are ☐ times 16 ounces.

☐ x 16 = ☐ ounces

Note taking

2b Look at the units in the picture.
How many ounces are there in 11 pounds?

1 lb = 16 oz

There are ☐ times 16 ounces.

☐ x 16 = ☐ ounces

Note taking

2c There are 12 inches in 1 foot.

How many inches are in 7 feet?

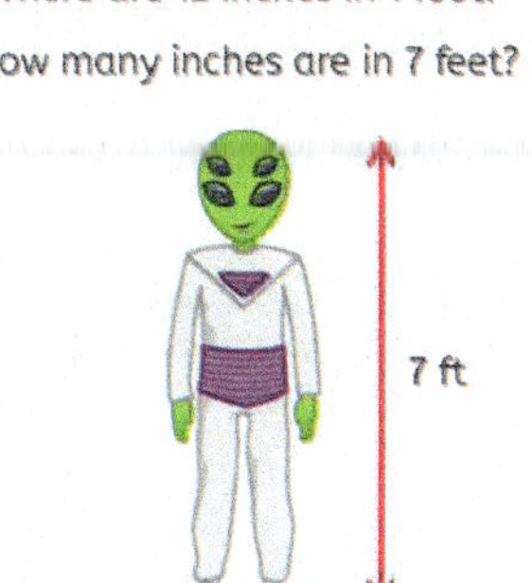

7 feet is ☐ × ☐ inches = ☐ inches

Note taking

2d

3 ft = 1 yd

How many feet (ft) are there in 9 yards (yd)?

- 9 ft
- 18 ft
- 27 ft
- 90 ft

Note taking

2e

1 lb = 16 oz

How many ounces (oz) are there in 6 pounds (lb)?

- 32 oz
- 36 oz
- 72 oz
- 96 oz

Note taking

2f

2,000 lb = 1 t

How many pounds (lb) are there in 10 tons (t)?

- 120 lb
- 4,000 lb
- 10,000 lb
- 20,000 lb

Note taking

2g

12 in. = 1 ft

How many inches (in.) are there in 12 feet (ft)?

- 24 in.
- 120 in.
- 144 in.
- 20,000 in.

Note taking

2h Hint

Convert feet to inches.

length in ft		in. in 1 ft		length in in.
5	×		=	

Note taking

2i Hint

Convert feet to inches.

length in ft		in. in 1 ft		length in in.
9	×		=	

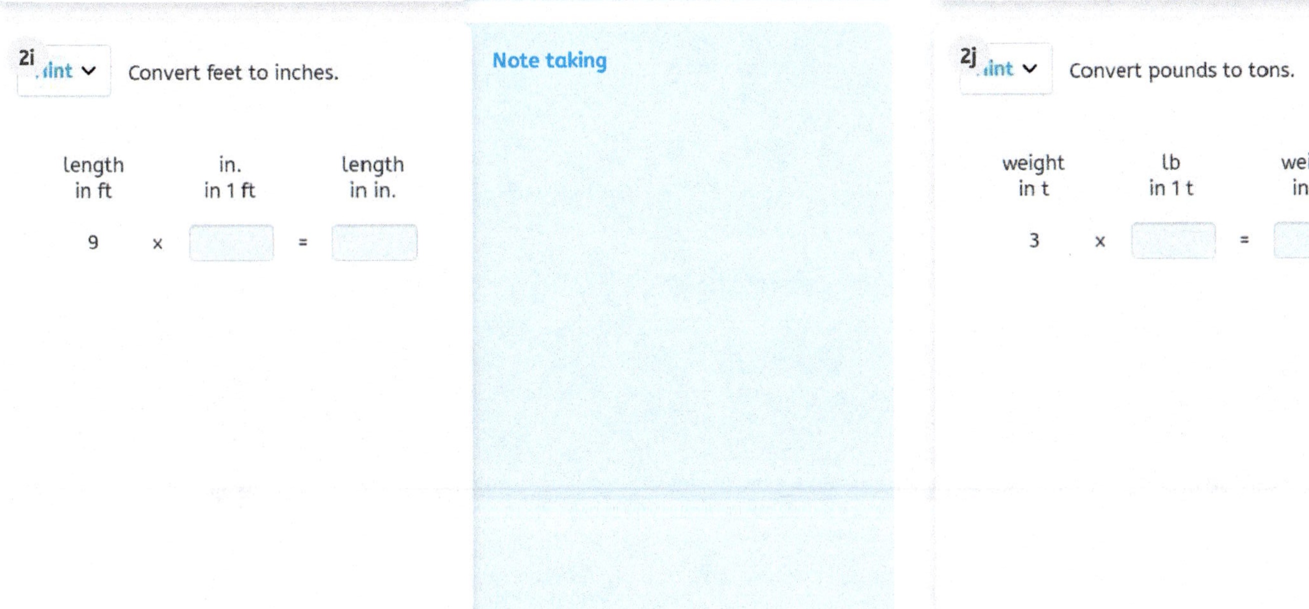

Note taking

2j Hint

Convert pounds to tons.

weight in t		lb in 1 t		weight in lb
3	×		=	

Note taking

2k

Complete the conversion table.

1 yd	**36 in.**
2 yd	☐ in.
3 yd	☐ in.
5 yd	☐ in.
10 yd	☐ in.

Note taking

2l

Complete the conversion table.

1 lb	☐ oz
2 lb	**32 oz**
4 lb	☐ oz
6 lb	☐ oz
8 lb	☐ oz

Note taking

2m

Complete the conversion table.

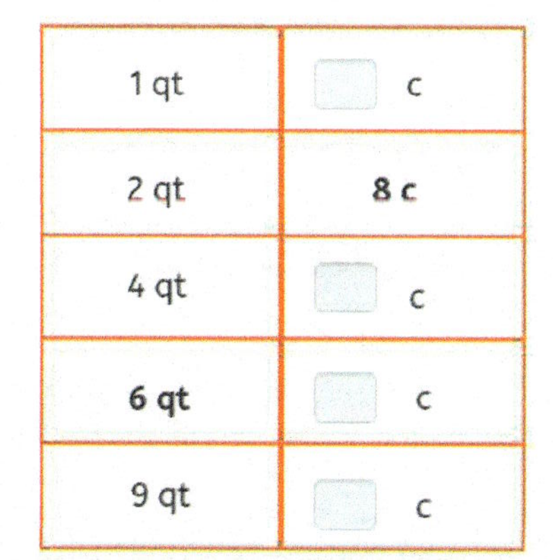

1 qt	☐ c
2 qt	**8 c**
4 qt	☐ c
6 qt	☐ c
9 qt	☐ c

Note taking

1a

Do you remember?

There are 16 oz in 1 lb.

How many oz are there in 3 lb?

☐ oz

Note taking

1b

Do you remember?

There are 4 c in 1 qt.

How many c are there in 10 qt?

☐ c

Note taking

1i

Rachel's dog weighs 29 kg.

Fiona's dog weighs 26 kg.

How much do the dogs weigh together in hg?

Steps

☐ hg

× 10 × 10 × 10

kg hg dag g

÷ 10 ÷ 10 ÷ 10

Note taking

1j

Each spoon weighs 7 g.

What do 12 spoons weigh in dag?

Steps

☐ dag

× 10 × 10 × 10

kg hg dag g

÷ 10 ÷ 10 ÷ 10

Note taking

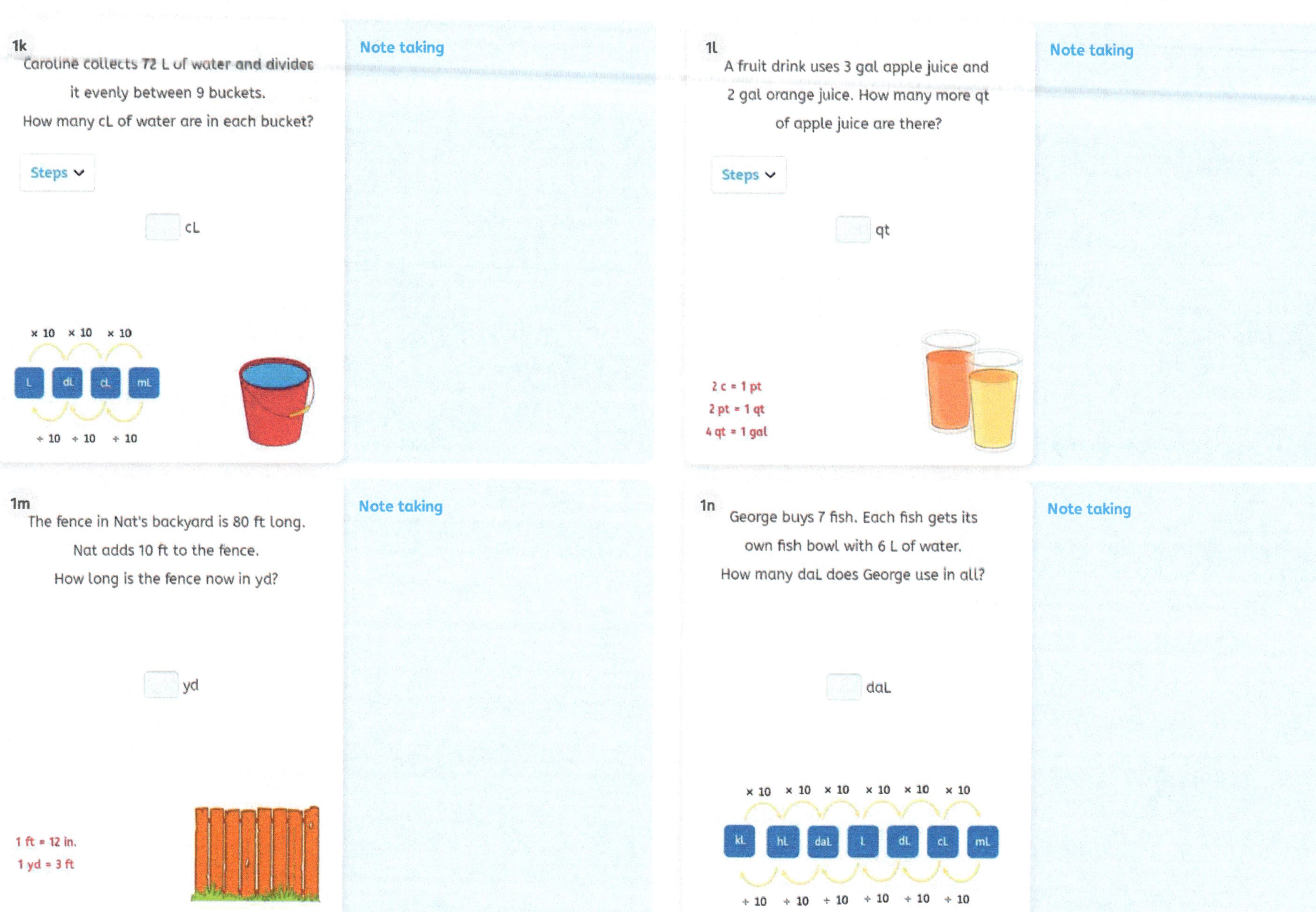

1k

Caroline collects 72 L of water and divides it evenly between 9 buckets.

How many cL of water are in each bucket?

Steps

☐ cL

Note taking

1l

A fruit drink uses 3 gal apple juice and 2 gal orange juice. How many more qt of apple juice are there?

Steps

☐ qt

2 c = 1 pt
2 pt = 1 qt
4 qt = 1 gal

Note taking

1m

The fence in Nat's backyard is 80 ft long.

Nat adds 10 ft to the fence.

How long is the fence now in yd?

☐ yd

1 ft = 12 in.
1 yd = 3 ft

Note taking

1n

George buys 7 fish. Each fish gets its own fish bowl with 6 L of water.

How many daL does George use in all?

☐ daL

Note taking

1o

How can Tina fill a large container with smaller containers of 200g, 600g, 800g each without exceeding a 12 kg weight limit?

Write down two solutions below:

Note taking

2a

Ben uses 15 L of water to wash the dishes and 6 L of water to water the plants.

How many daL of water does Ben use?

Steps

☐ daL

× 10 × 10 × 10

kL hL daL L

÷ 10 ÷ 10 ÷ 10

Note taking

2b

Anna needs 3 quarts of stock and 1 quart of water to make soup. How many cups of stock and water does she need?

Steps

☐ cups

2 c = 1 pt
2 pt = 1 qt
4 qt = 1 gal

Note taking

2c

A baby elephant weighs 200 pounds at birth. How many tons do 10 baby elephants weigh?

Steps

☐ T

16 oz = 1 lb
2000 lb = 1 T

Note taking

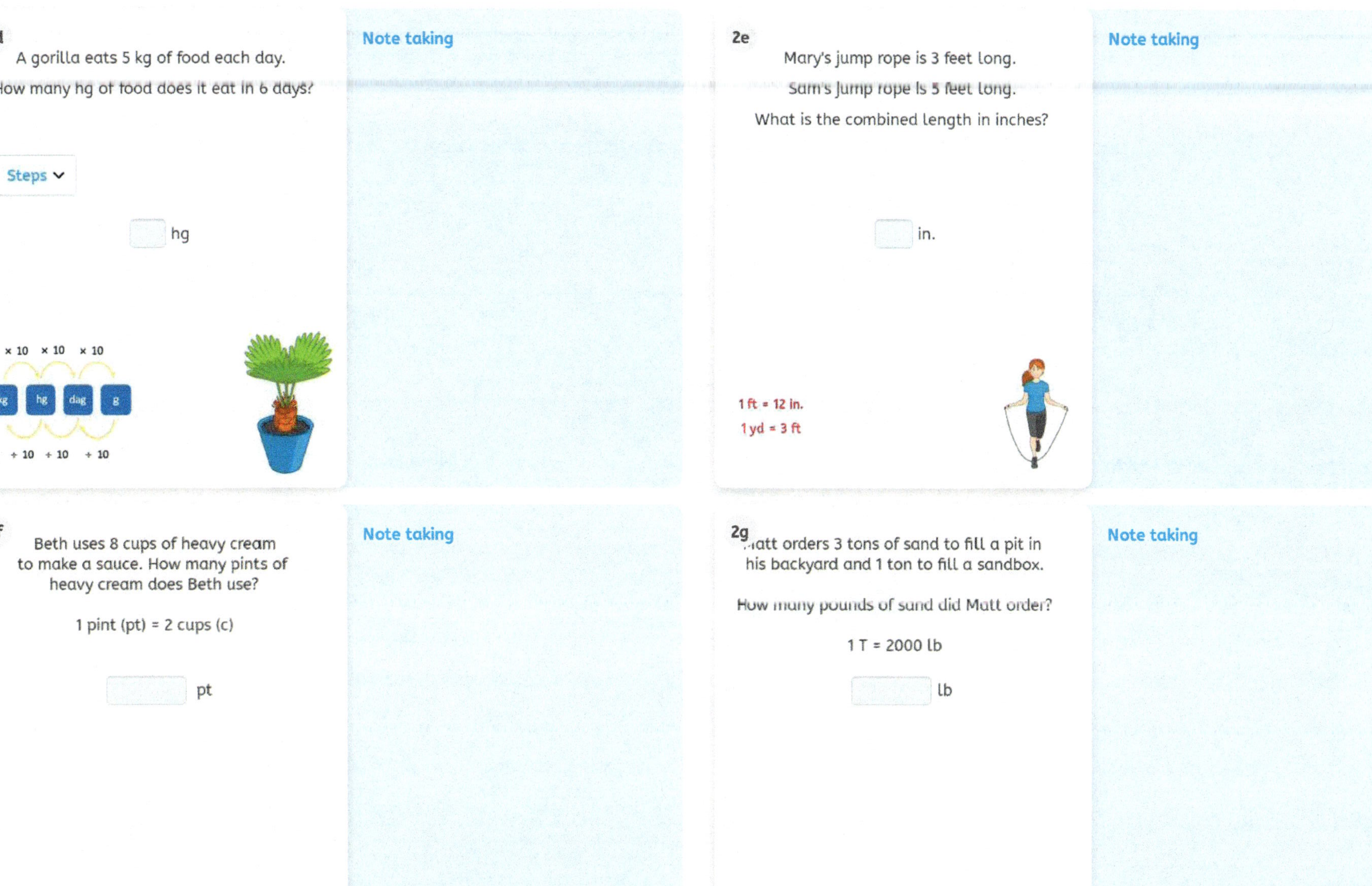

2d

A gorilla eats 5 kg of food each day.

How many hg of food does it eat in 6 days?

Steps

☐ hg

Note taking

2e

Mary's jump rope is 3 feet long.

Sam's jump rope is 5 feet long.

What is the combined length in inches?

☐ in.

1 ft = 12 in.

1 yd = 3 ft

Note taking

2f

Beth uses 8 cups of heavy cream to make a sauce. How many pints of heavy cream does Beth use?

1 pint (pt) = 2 cups (c)

☐ pt

Note taking

2g

Matt orders 3 tons of sand to fill a pit in his backyard and 1 ton to fill a sandbox.

How many pounds of sand did Matt order?

1 T = 2000 lb

☐ lb

Note taking

2h

Tom buys 15 pounds of food for his dog and 4 pounds of food for his cat.
How many ounces of food does Tom buy?

1 pound (lb) = 16 ounces (oz)

___ oz

Note taking

2i

Stacey races 50 m in her first race and 90 m in her second race.
How many dm does Stacey race?

___ dm

× 10 × 10 × 10
m dm cm mm
÷ 10 ÷ 10 ÷ 10

Note taking

2j

One kitten weighs 2 kg.
Another kitten weighs 4 kg.
What is their combined weight in hg?

___ hg

× 10 × 10 × 10
kg hg dag g
÷ 10 ÷ 10 ÷ 10

Note taking

2k

How can Timothy fill a large container with smaller containers of 200g, 300g, 500g each without exceeding a 9 kg weight limit?

Write down two solutions below:

Note taking

1a Do you remember?

How many square tiles does it take to cover the front of the book?

[ ] square tiles

Note taking

1b Do you remember?

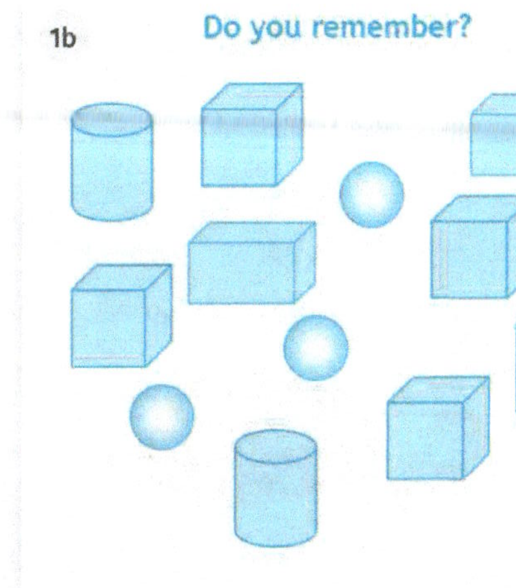

How many cubes are there?

[ ] cubes.

Note taking

1n These cubes fit exactly in a box without any gaps or overlap.

Each cube is 1 cm^3.

What is the volume of box?

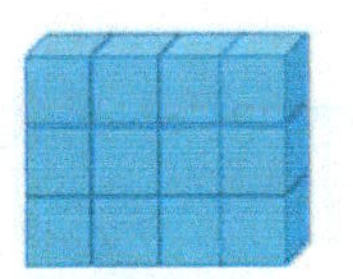

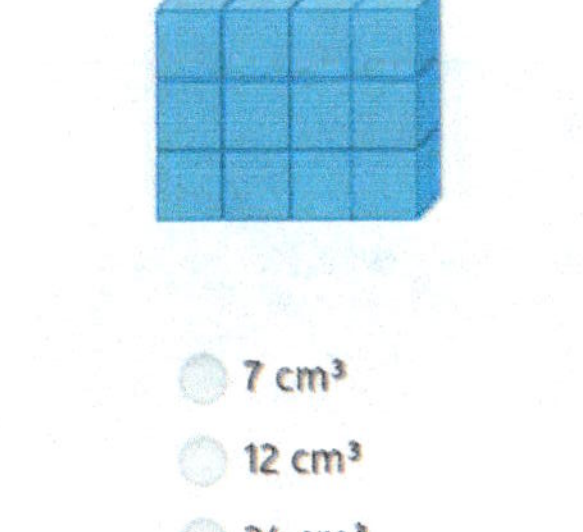

- 7 cm^3
- 12 cm^3
- 24 cm^3

Note taking

1o Each cube is 1 $in.^3$.

What is the volume of the figure?

Tip

- 16 $in.^3$
- 20 $in.^3$
- 32 $in.^3$

Note taking

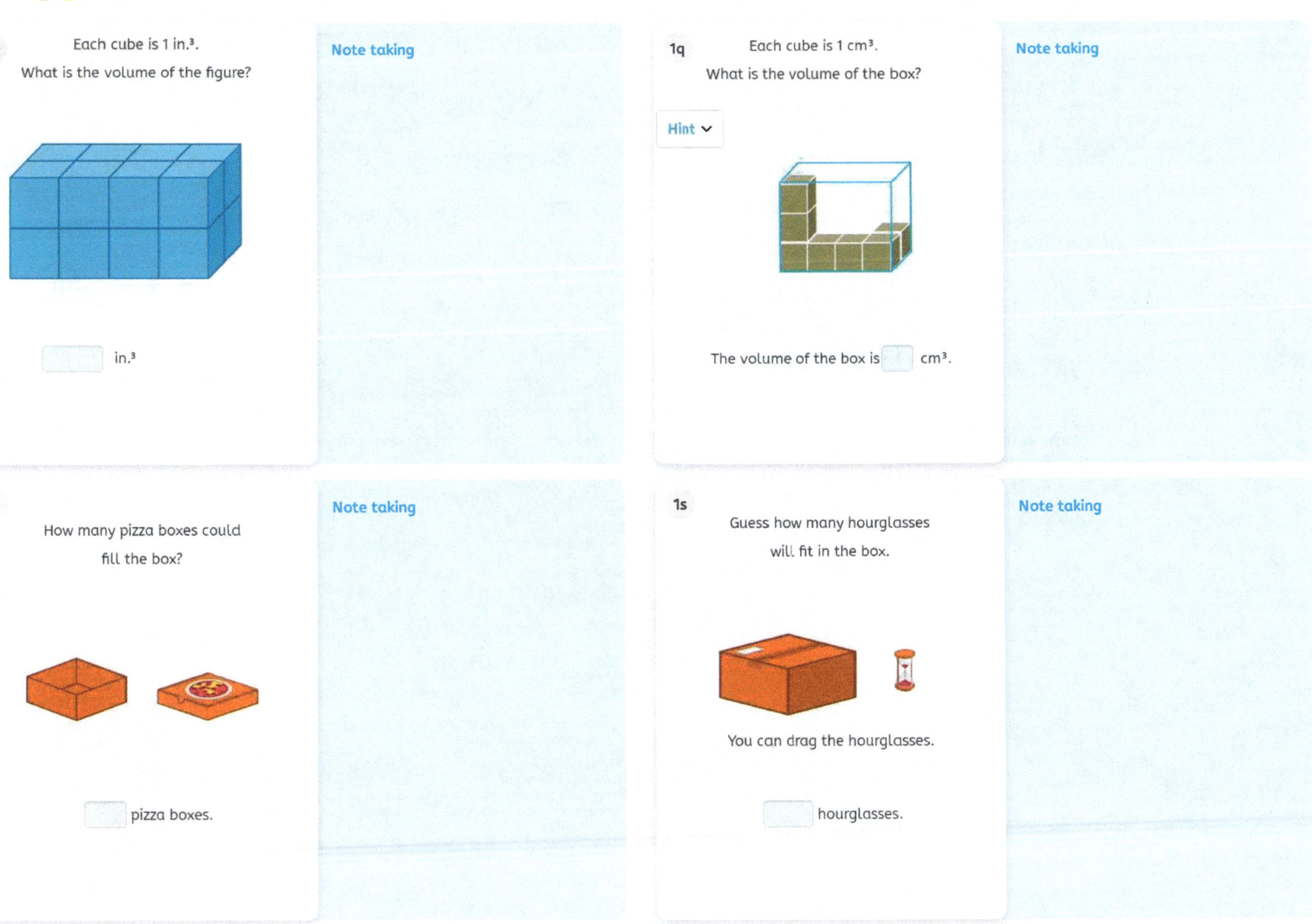

1p Each cube is 1 in.^{3}.

What is the volume of the figure?

____ in.^{3}

Note taking

1q Each cube is 1 cm^3.

What is the volume of the box?

Hint

The volume of the box is ____ cm^3.

Note taking

1r How many pizza boxes could fill the box?

____ pizza boxes.

Note taking

1s Guess how many hourglasses will fit in the box.

You can drag the hourglasses.

____ hourglasses.

Note taking

2a These cubes fit exactly in a box without any gaps or overlap.

Each cube is 1 cm^3.

What is the volume of box?

- 8 cm^3
- 15 cm^3
- 24 cm^3

Note taking

2b These cubes fit exactly in a box without any gaps or overlap.

Each cube is 1 cm^3.

What is the volume of box?

- 8 cm^3
- 15 cm^3
- 24 cm^3

Note taking

2c Each cube is 1 $in.^3$.

What is the volume of the figure?

- 16 $in.^3$
- 24 $in.^3$
- 36 $in.^3$

Note taking

2d Each cube is 1 $in.^3$.

What is the volume of the figure?

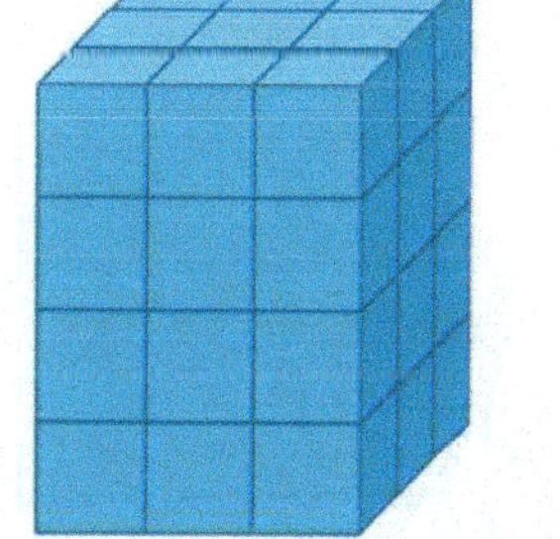

- 16 $in.^3$
- 24 $in.^3$
- 36 $in.^3$

Note taking

2e Each cube is 1 in.^{3}.

What is the volume of the figure?

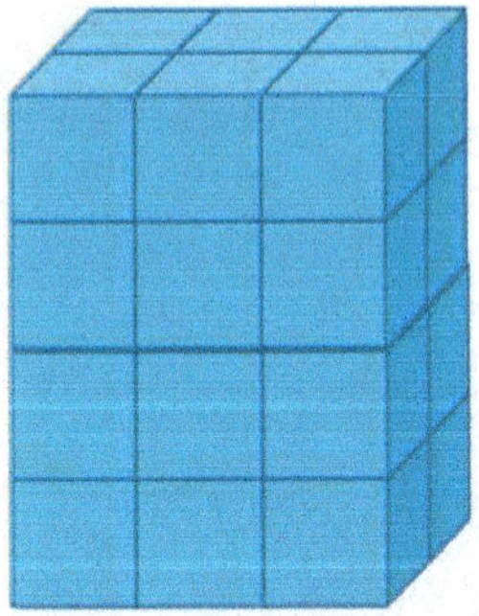

☐ in.^{3}

Note taking

2f Each cube is 1 in.^{3}.

What is the volume of the figure?

☐ in.^{3}

Note taking

2g Each cube is 1 in.^{3}.

What is the volume of the figure?

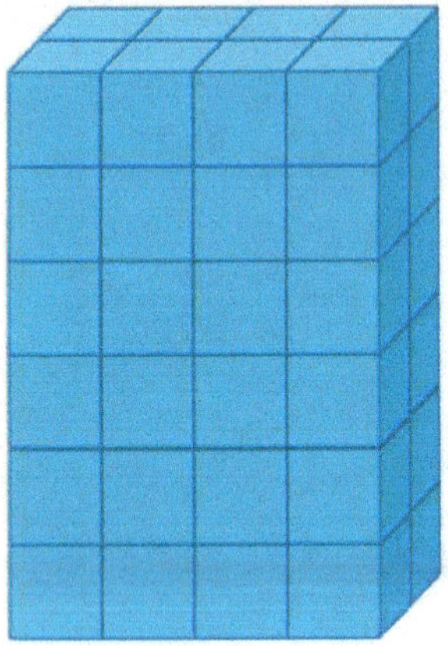

☐ in.^{3}

Note taking

2h Each cube is 1 in.^{3}.

What is the volume of the figure?

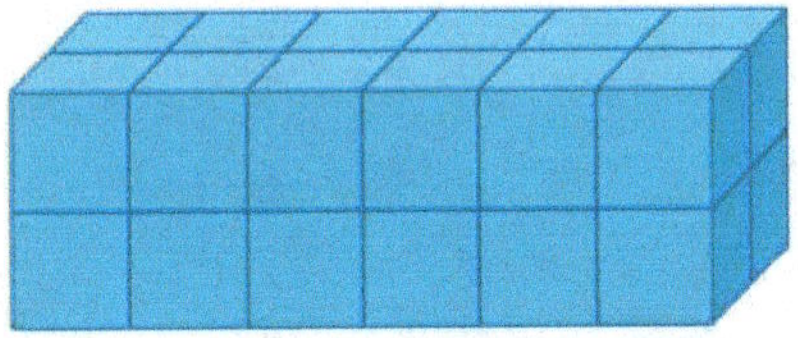

☐ in.^{3}

Note taking

2i Each cube is 1 cm^3.

What is the volume of the box?

Hint

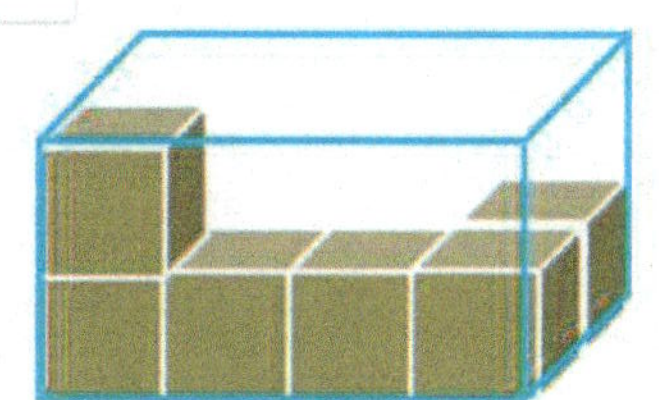

The volume of the box is ...

- 6 cm^3
- 10 cm^3
- 16 cm^3

Note taking

2j Each cube is 1 $in.^3$.

What is the volume of the box?

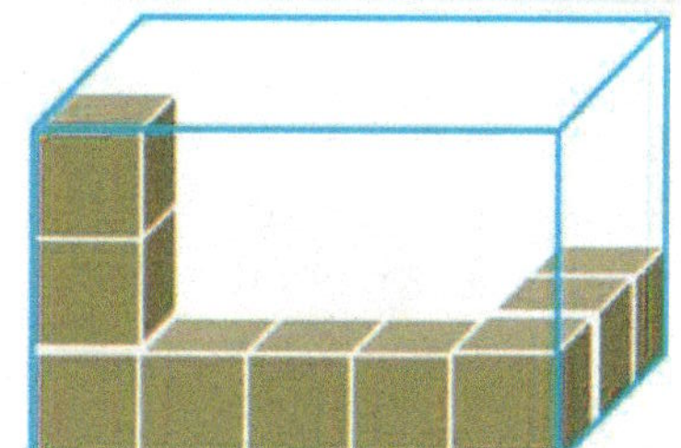
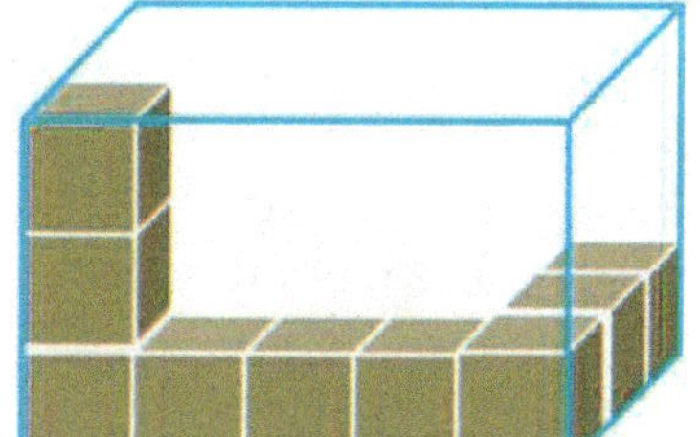

The volume of the box is ___ $in.^3$.

Note taking

2k Guess how many hourglasses will fit in the box.

You can drag the hourglasses.

___ hourglasses.

Note taking

1a Do you remember?

☐ x 3 = ☐

Note taking

1b Do you remember?

4 x 7 x 2 = ...

- 13
- 14
- 28
- 56

Note taking

1c Do you remember?

These cubes fill a box without gaps or overlap.

What is the volume of the box?

☐ cubic units

Note taking

1l

Each cube is 1 cm^3.

The base layer has 8 cubes.

What is the volume of the figure?

1 cm
1 cm 1 cm

8 x ☐ = 40 cm^3

Note taking

1m

Each cube is 1 in.³.

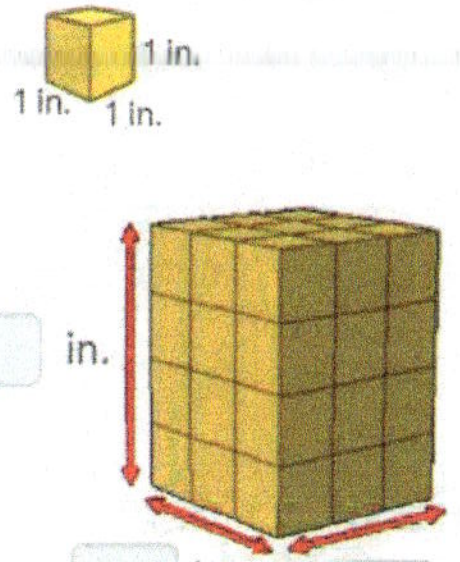

1 in. 1 in. 1 in.

___ in.

___ in. ___ in.

What is the volume of the figure?

___ in.³

Note taking

1n

What is the volume?

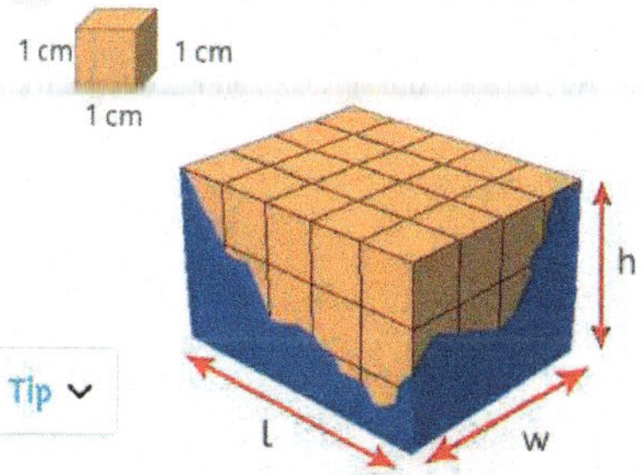

1 cm 1 cm 1 cm

Tip

h

l w

length x width x height

5 cm x 4 cm x ___ cm

___ cm³

Note taking

1o

Each cube is 1 cm³.

The base layer has 10 cubes.

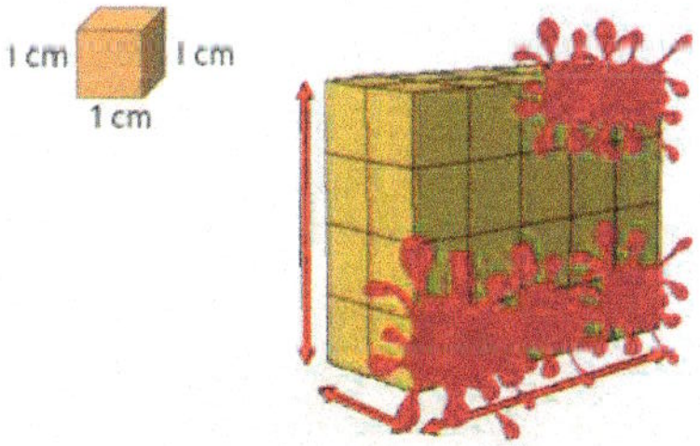

1 cm 1 cm 1 cm

What is the volume of the figure?

10 x 4 = ___

The volume is ___ cm³.

Note taking

2a

Each cube is 1 cm³.

What is the volume of the figure?

___ cm x ___ cm x ___ cm = 48 cm³

Note taking

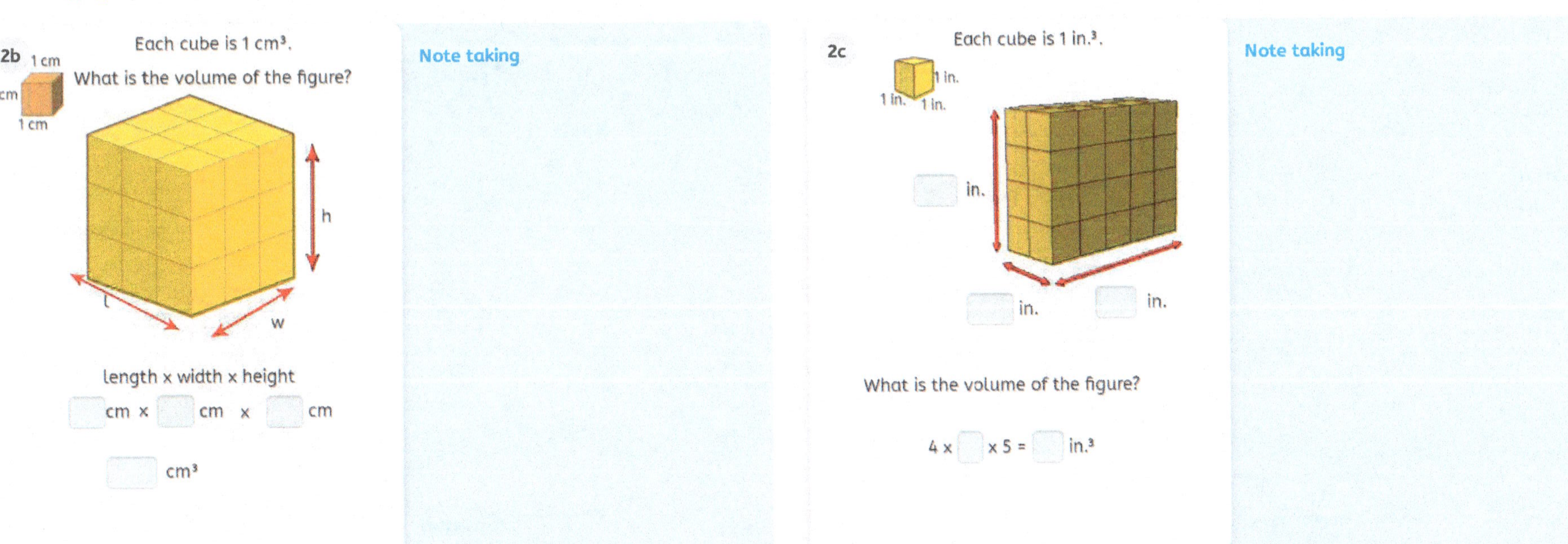

2b

Each cube is 1 cm³.

What is the volume of the figure?

length x width x height

[ ] cm x [ ] cm x [ ] cm

[ ] cm³

Note taking

2c

Each cube is 1 in.³.

[ ] in.

[ ] in. [ ] in.

What is the volume of the figure?

4 x [ ] x 5 = [ ] in.³

Note taking

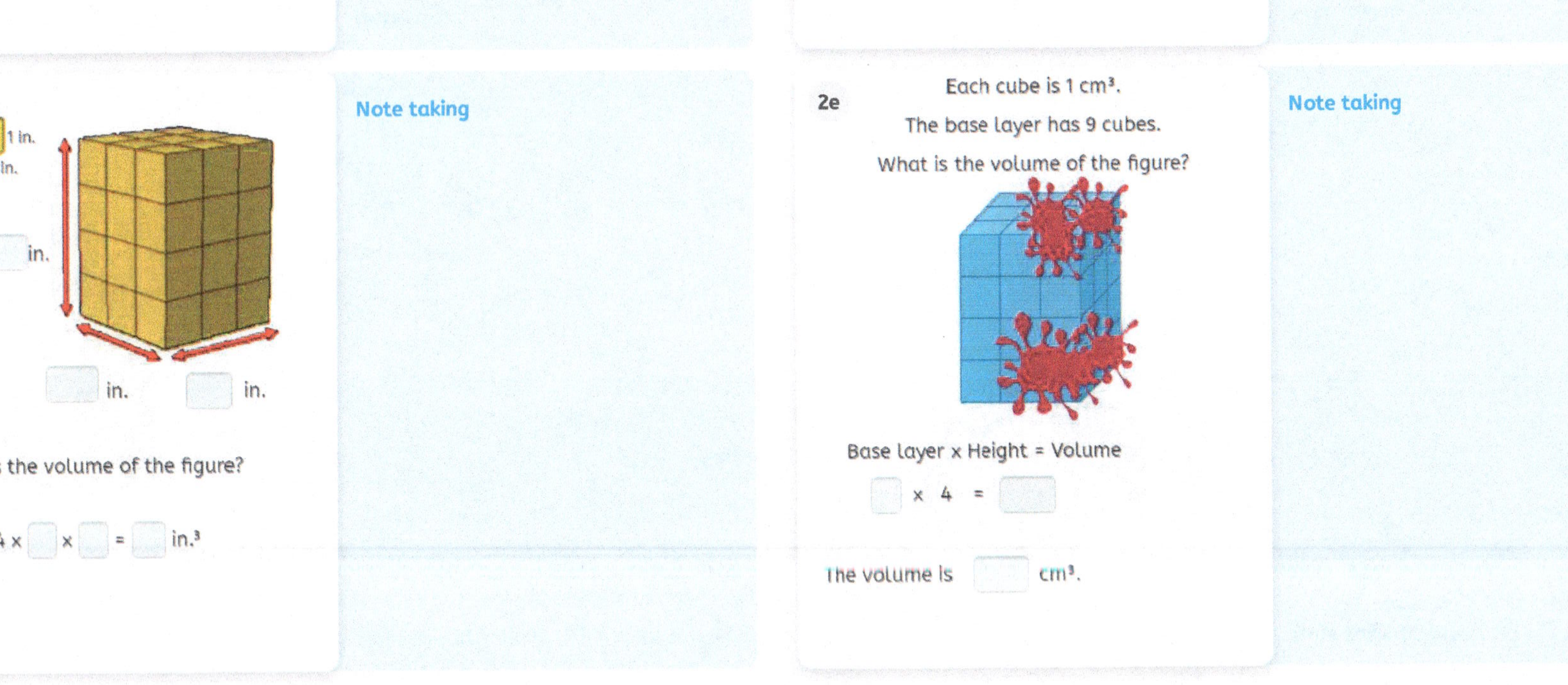

2d

[ ] in.

[ ] in. [ ] in.

What is the volume of the figure?

4 x [ ] x [ ] = [ ] in.³

Note taking

2e

Each cube is 1 cm³.

The base layer has 9 cubes.

What is the volume of the figure?

Base layer x Height = Volume

[ ] x 4 = [ ]

The volume is [ ] cm³.

Note taking

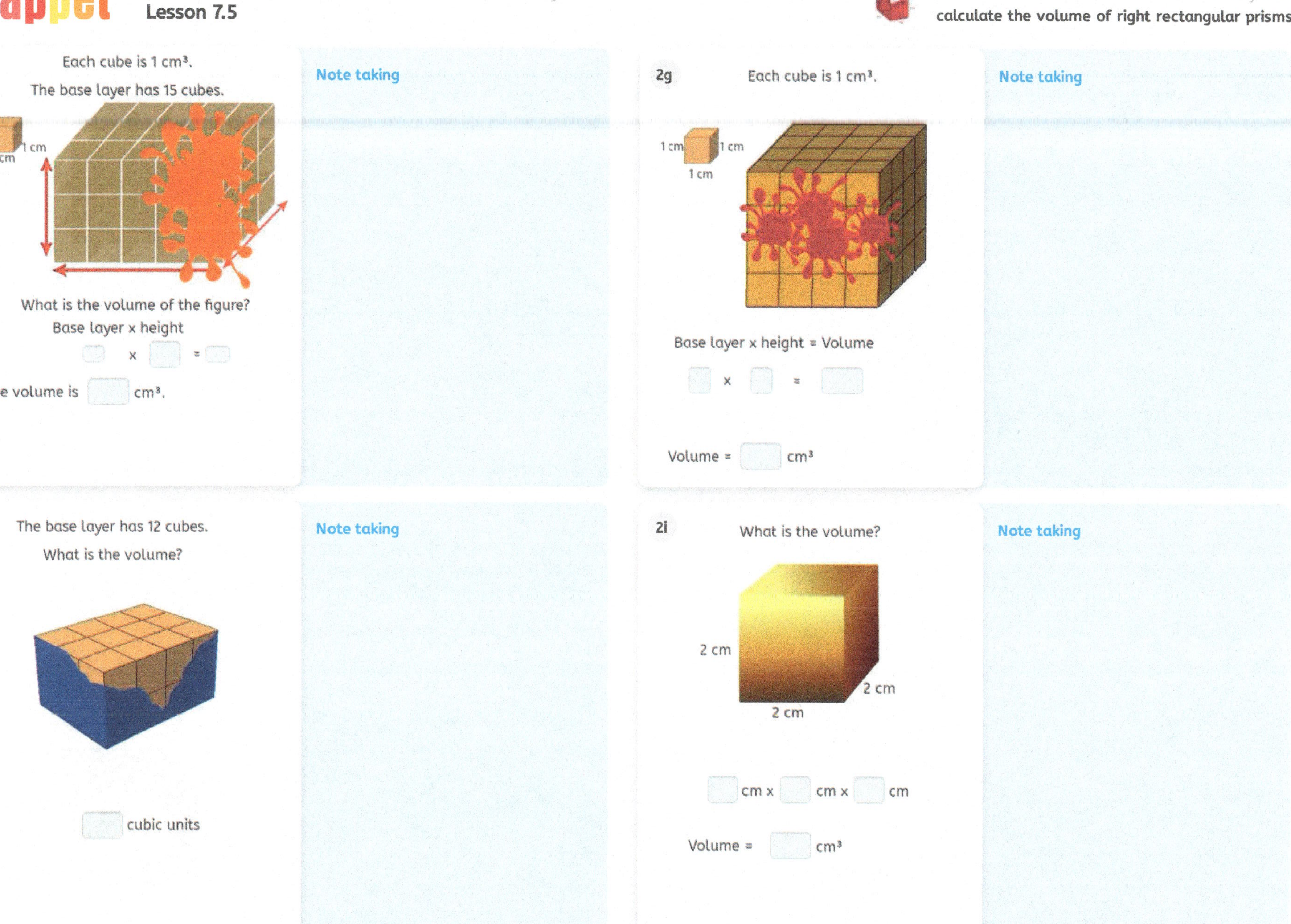

2f Each cube is 1 cm³.

The base layer has 15 cubes.

What is the volume of the figure?

Base layer x height

☐ x ☐ = ☐

The volume is ☐ cm³.

Note taking

2g Each cube is 1 cm³.

Base layer x height = Volume

☐ x ☐ = ☐

Volume = ☐ cm³

Note taking

2h The base layer has 12 cubes.

What is the volume?

☐ cubic units

Note taking

2i What is the volume?

☐ cm x ☐ cm x ☐ cm

Volume = ☐ cm³

Note taking

2j What is the volume?

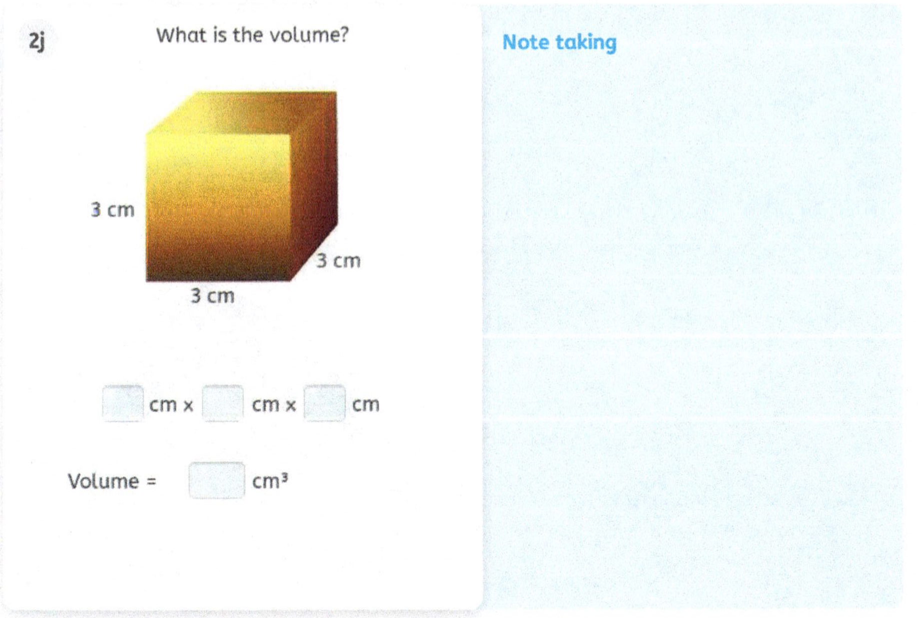

☐ cm x ☐ cm x ☐ cm

Volume = ☐ cm^3

Note taking

1a **Do you remember?**

Each cube is 1 in³.
What are the dimensions of the figure?

height = ☐ in.

length = ☐ in. width = ☐ in.

Note taking

1b **Do you remember?**

Each side of one cube is 1 inch.
What is the volume of the figure?

- ◯ 8 in.³
- ◯ 15 in.³
- ◯ 20 in.³

Note taking

1c **Do you remember?**

$3 \times 2 \times 5 =$ ☐

Note taking

1j Each side of a cube is 1 in.

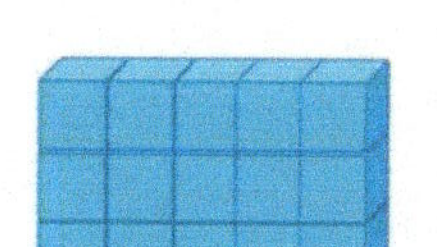

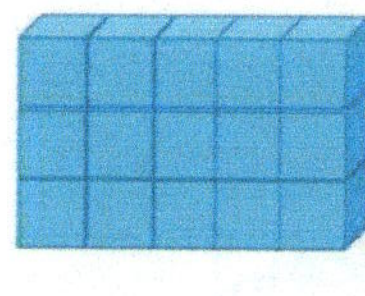

The volume of each figure is:

$\ell \times w \times h =$ volume

☐ in. × ☐ in. × ☐ in. = ☐ in.³

Note taking

1k What is the volume of the suitcase?

20 in.
8 in.
10 in.

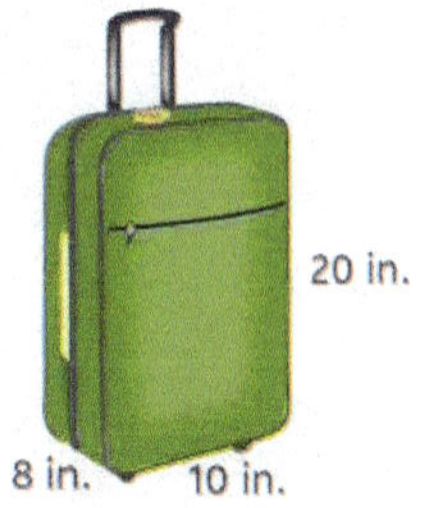

Volume = length x width x height

Volume = ☐ x 8 x 20

The volume of the suitcase is ☐ in.³.

Note taking

1m What is the volume of the box?

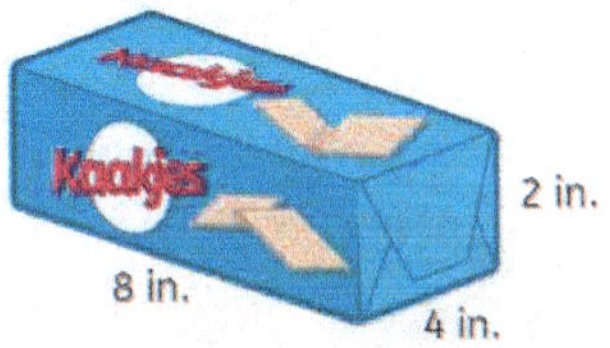

2 in.
8 in.
4 in.

Tip

The volume of the box is ☐ in.³.

Note taking

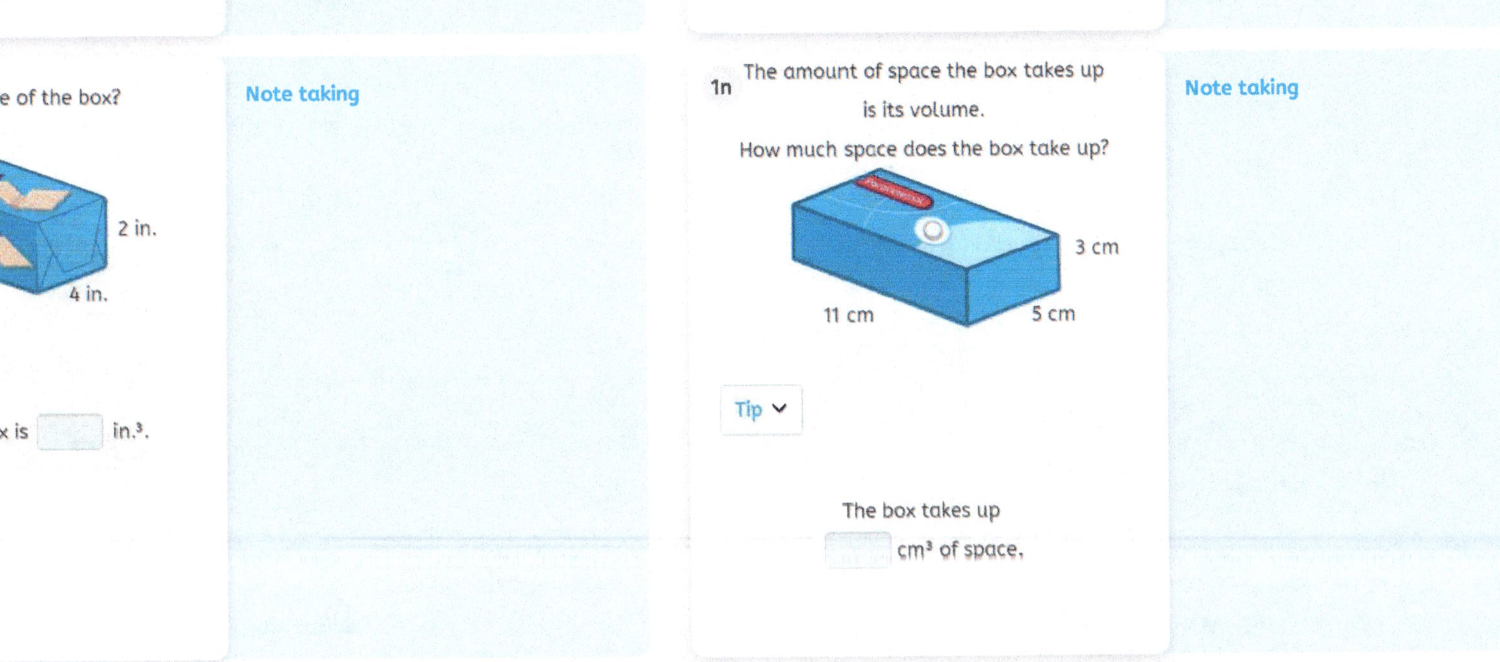

1l

10 cm
9 cm
8 cm

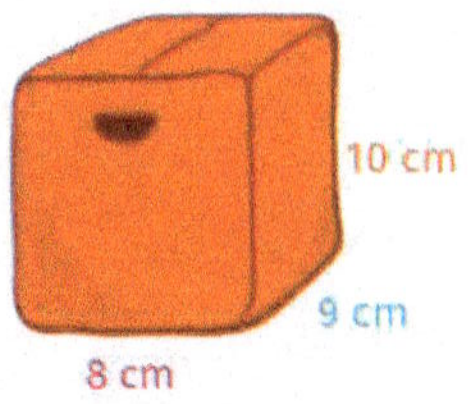

The volume of the box is:

ℓ x w x h = V

☐ x ☐ x ☐ = ☐ cm³

Note taking

1n The amount of space the box takes up is its volume.

How much space does the box take up?

3 cm
11 cm
5 cm

Tip

The box takes up ☐ cm³ of space.

Note taking

1o

The base area of this package is 60 cm^2.

How much butter can the package hold?

3 cm

Tip

The package holds ☐ cm^3 of butter.

Note taking

1p

Tip

The base area of the building is 25 m^2.

The height of the building is 5 m.

What is the volume?

The volume is ☐ m^3.

Note taking

1q

The base area of this room is 64 ft^2.

The height of the room is 10 ft.

What volume of air will this room hold?

Tip

The room holds ☐ ft^3 of air.

Note taking

2a

What is the volume?

8 in.

3 in.

12 in.

length x width x height

12 in. x ☐ in. x 8 in.

☐ $in.^3$

Note taking

2b What is the volume of the box?

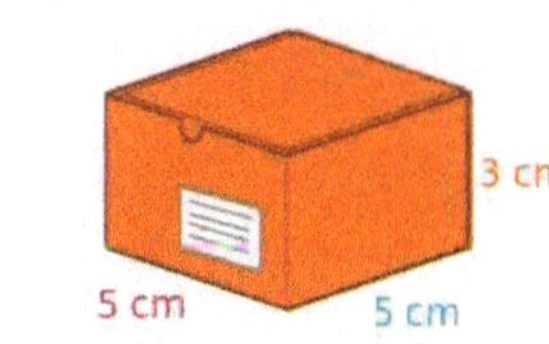

The volume of the box can be found using:

ℓ x w x h = volume

[ ] cm x [ ] cm x [ ] cm = 75 cm³

Note taking

2c What is the volume of the cake pan?

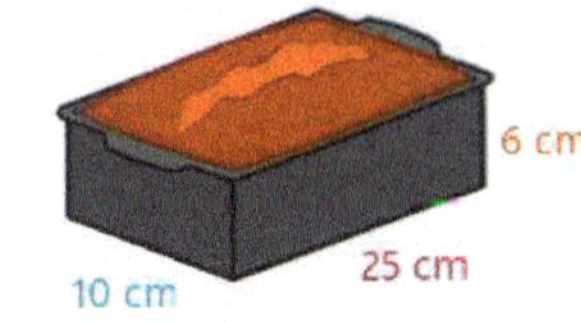

The volume of this cake pan is:

ℓ x w x h = volume

[ ] cm x [ ] cm x [ ] cm = [ ] cm³

Note taking

2d How much space is inside the box?

Hint

5 cm
8 cm
6 cm

[ ] cm³

Note taking

2e

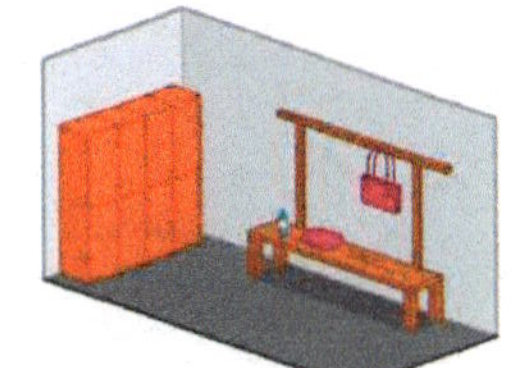

The base area of this locker room is 10 yd².

The height is 3 yd.

How much air can the locker room hold?

Tip

The locker room can hold [ ] yd³ of air.

Note taking

2f Joanna makes dot cubes using foam.
Each cube fits exactly in a box that has a height of 5 in. and a base area of 25 in.^{2}.
What is the volume of the box?

Tip

The volume of the box is ☐ in.^{3}.

Note taking

2g

10 ft

The base area of this storage container is 60 ft^2.
How much space is in the storage container?

Tip

The storage container has ☐ ft^3 of space.

Note taking

2h A sandbox is 6 feet long, 4 feet wide, and 2 feet high.
How much sand does it take to fill the sandbox?

☐ ft^3 of sand

Note taking

2i
A pool is in the shape of a rectangular prism.
The base area of the pool is 40 m^2.
The pool is 3 m deep.
How much water can the pool hold?

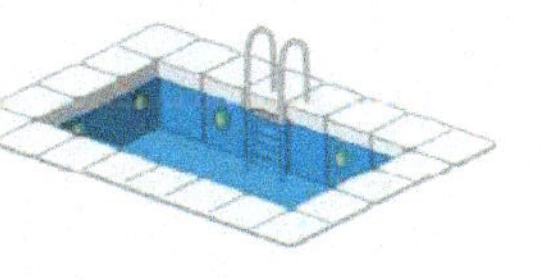

☐ m^3 of water

Note taking

2j

A gift bag in the shape of a rectangular prism is 12 in. long, 4 in. wide, and 12 in. high. What is the volume of the bag?

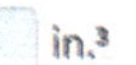
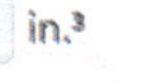
☐ $in.^3$

Note taking

1a Do you remember?

6 cm
3 cm
4 cm

The volume of the box is:

ℓ × w × h = V

☐ × ☐ × ☐ = ☐ cm³

Note taking

1b Do you remember?

Each side of a cube is 1 in.

What is the volume of the figure?

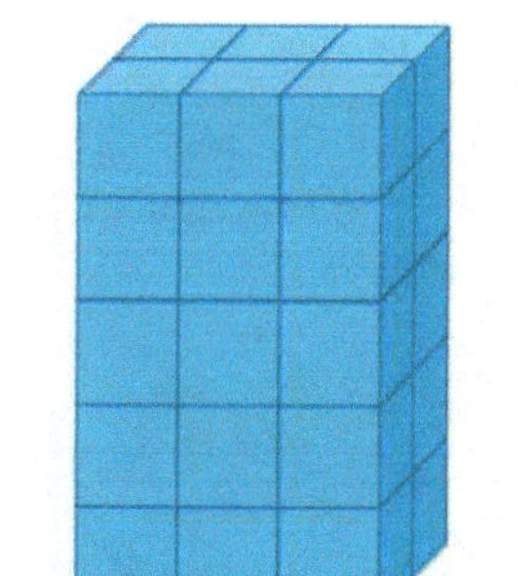
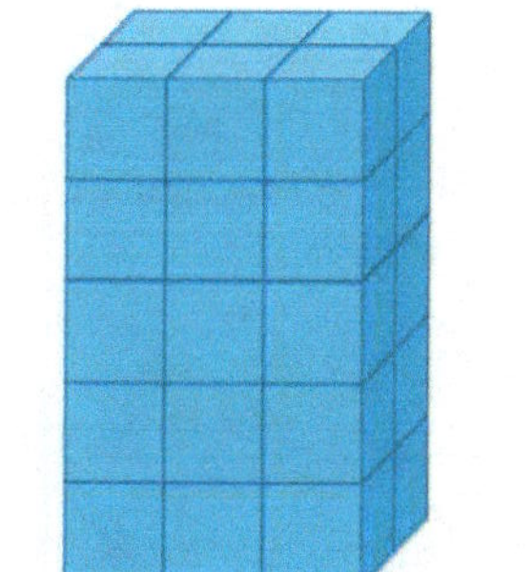

Volume = *Base* x *height*

The volume is ☐ x 5 = ☐ in.³

Note taking

1k Find the volume of the entire figure.

6 in.
4 in.
4 in.
6 in.
B
A
3 in.
3 in.
10 in.

Volume of A: ☐ x 4 x 3 = ☐ in.³

Volume of B: ☐ x 6 x 3 = ☐ in.³

Total volume: 48 + ☐ = ☐ in.³

Note taking

1l The two fish tanks are the same size. How much water does it take in all to fill them?

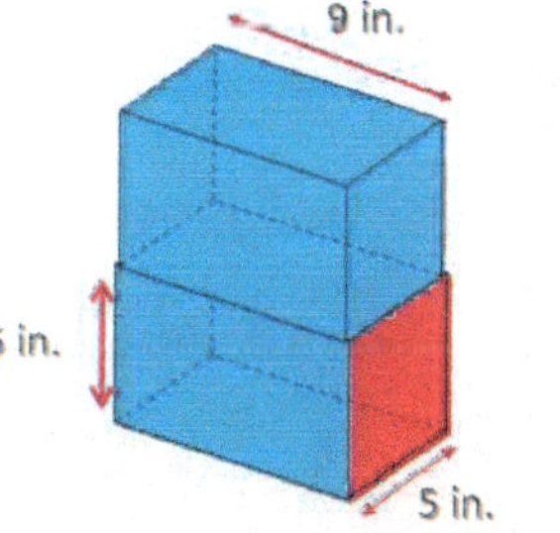
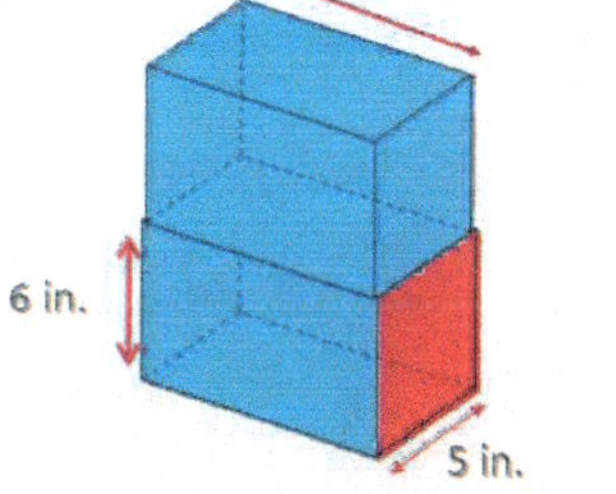

Volume of each fish tank:

9 x ☐ x 6 = ☐ in.³

Total Volume: ☐ + ☐ = ☐ in.³

Note taking

1m What is the volume of the figure?

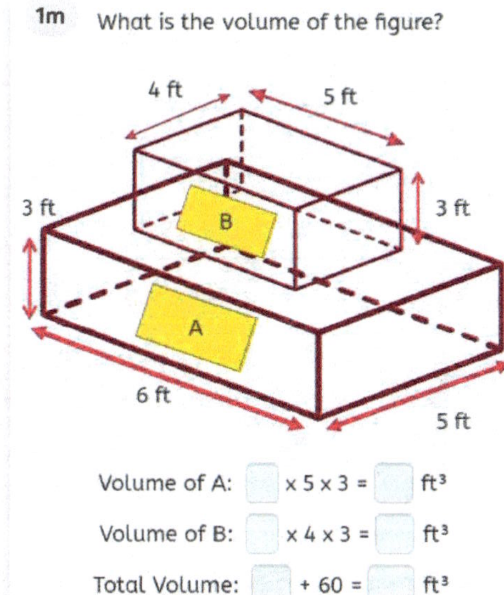

Volume of A: ☐ x 5 x 3 = ☐ ft^3

Volume of B: ☐ x 4 x 3 = ☐ ft^3

Total Volume: ☐ + 60 = ☐ ft^3

Note taking

1n What is the volume of the figure?

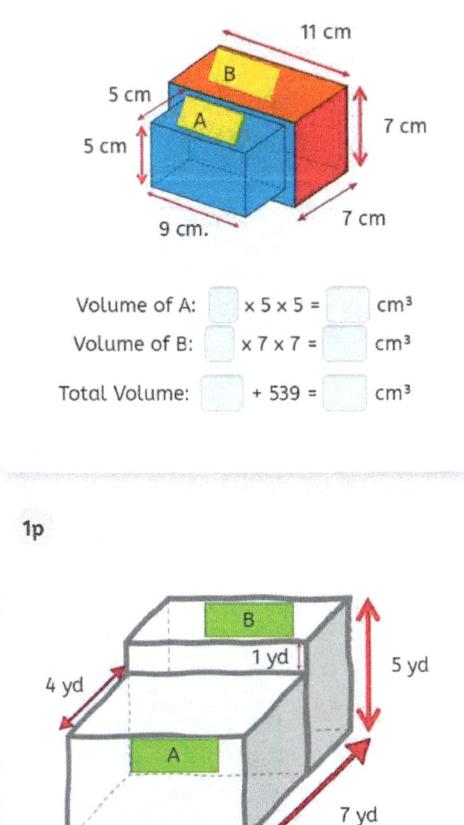

Volume of A: ☐ x 5 x 5 = ☐ cm^3

Volume of B: ☐ x 7 x 7 = ☐ cm^3

Total Volume: ☐ + 539 = ☐ cm^3

Note taking

1o ·culpture is made with two identical cubes. The cubes will be filled with concrete. What volume of concrete is needed?

- 8 ft^3
- 16 ft^3
- 24 ft^3

Note taking

1p

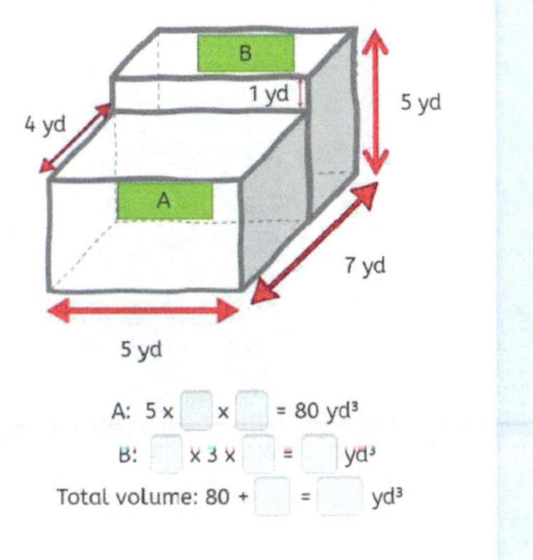

A: 5 x ☐ x ☐ = 80 yd^3

B: ☐ x 3 x ☐ = ☐ yd^3

Total volume: 80 + ☐ = ☐ yd^3

Note taking

2a Find the missing dimension and the volume of the entire figure.

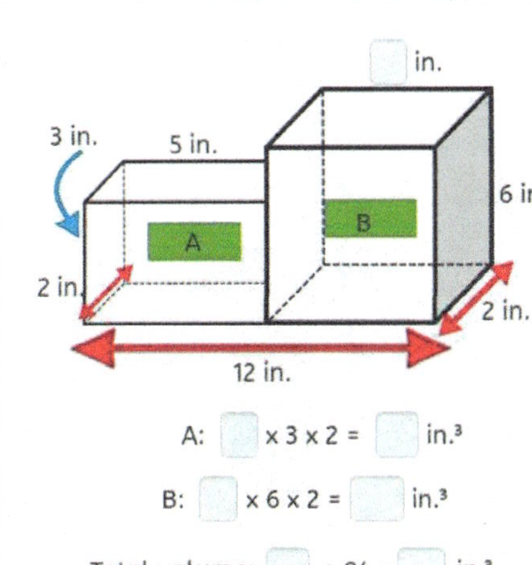

A: ___ x 3 x 2 = ___ in.³

B: ___ x 6 x 2 = ___ in.³

Total volume: ___ + 84 = ___ in.³

Note taking

2b Find the volume of the entire figure.

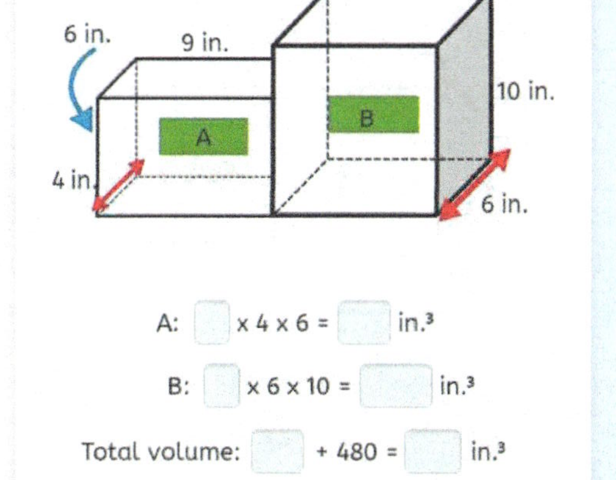

A: ___ x 4 x 6 = ___ in.³

B: ___ x 6 x 10 = ___ in.³

Total volume: ___ + 480 = ___ in.³

Note taking

2c What is the volume of the figure?

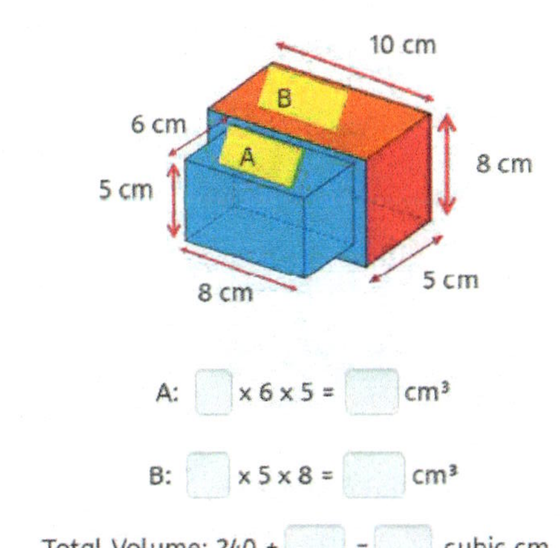

A: ___ x 6 x 5 = ___ cm³

B: ___ x 5 x 8 = ___ cm³

Total Volume: 240 + ___ = ___ cubic cm

Note taking

2d

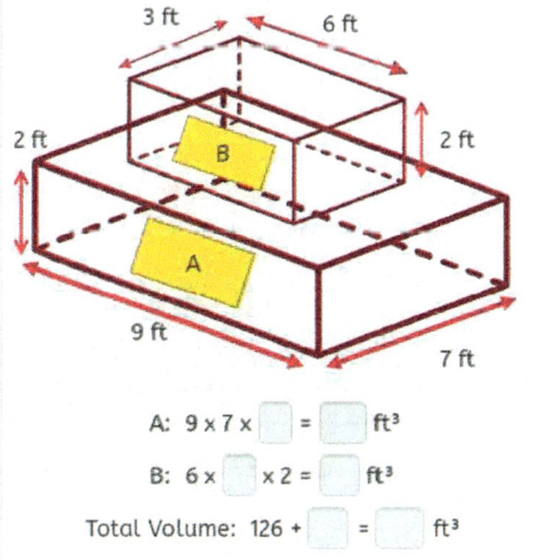

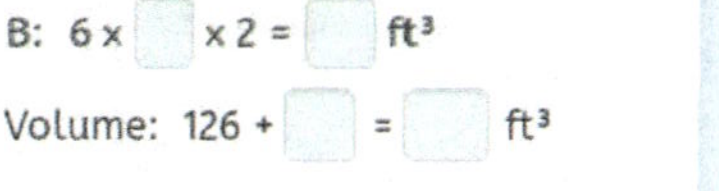

A: 9 x 7 x ___ = ___ ft³

B: 6 x ___ x 2 = ___ ft³

Total Volume: 126 + ___ = ___ ft³

Note taking

2e The two fish tanks are the same size. How much water does it take in all to fill them?

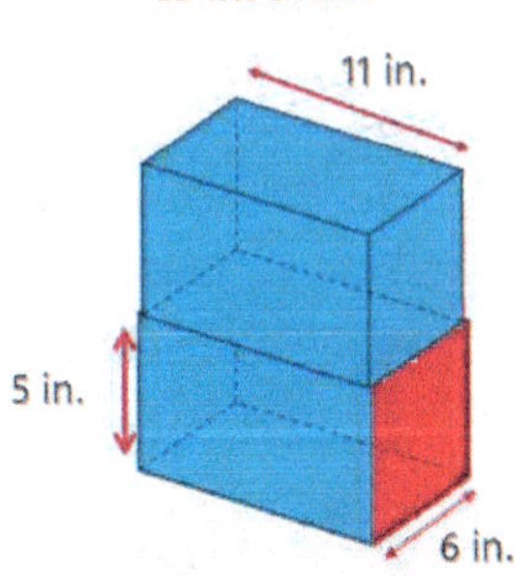

Volume of each fish tank:

11 x ___ x 5 = ___ in.³

Total Volume: ___ + ___ = ___ in.³

Note taking

2f

A: 6 x 5 x ___ = ___ yd³

B: ___ x 3 x ___ = ___ yd³

Total volume: ___ + 108 = ___ yd³

Note taking

2g The width of all parts of the structure is 4 in. What is the volume?

A: 6 x 4 x ___ = ___ in.³

B: 7 x ___ x 5 = ___ in.³

Total Volume: ___ + 140 = ___ in.³

Note taking

2h Find the missing dimension and the volume of the entire figure.

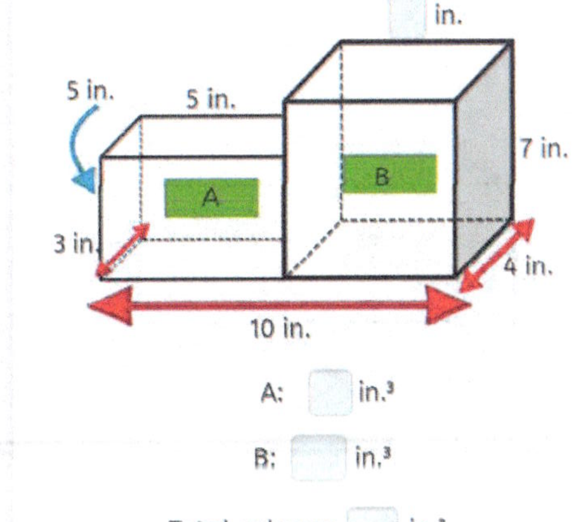

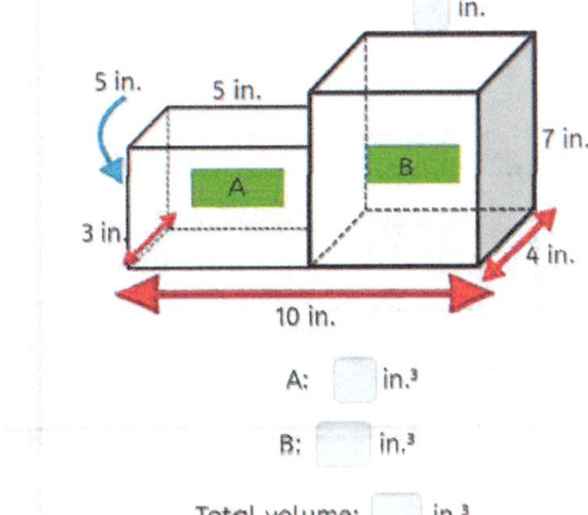

A: ___ in.³

B: ___ in.³

Total volume: ___ in.³

Note taking

2i ·culpture is made with two identical cubes.
The cubes will be filled with concrete.
What volume of concrete is needed?

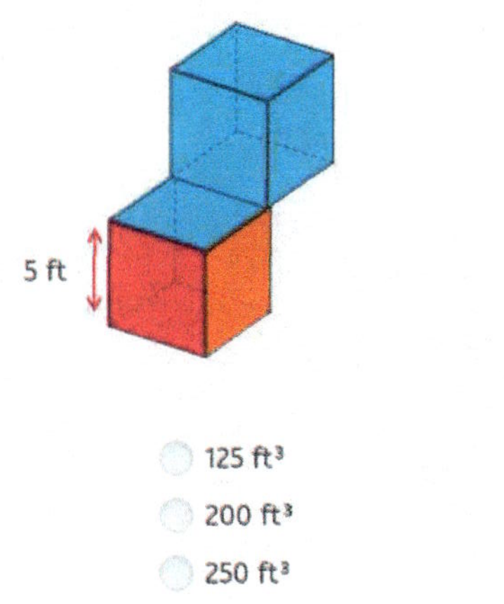

- 125 ft^3
- 200 ft^3
- 250 ft^3

Note taking

2j What is the volume of figure?

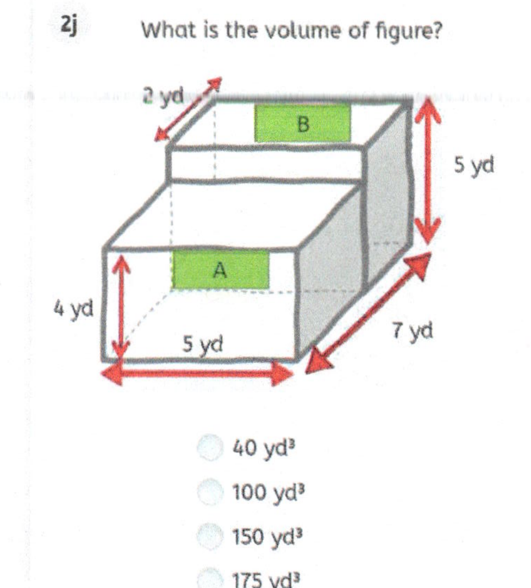

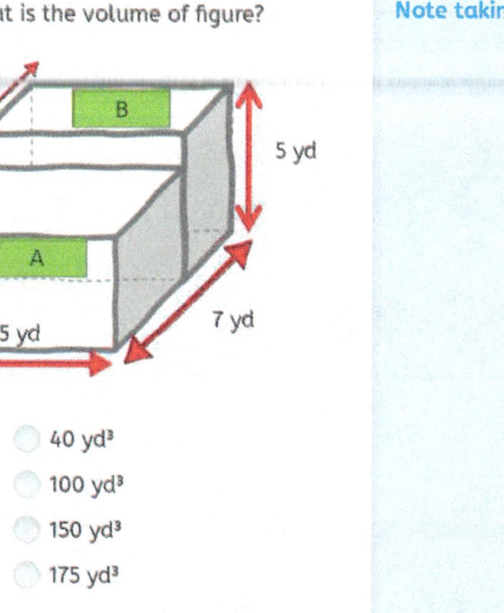

- 40 yd^3
- 100 yd^3
- 150 yd^3
- 175 yd^3

Note taking

2k T·· figure below represents the top portion of a concrete patio that will be 2 ft deep all around.
What is the volume of concrete that will be needed to create the patio?

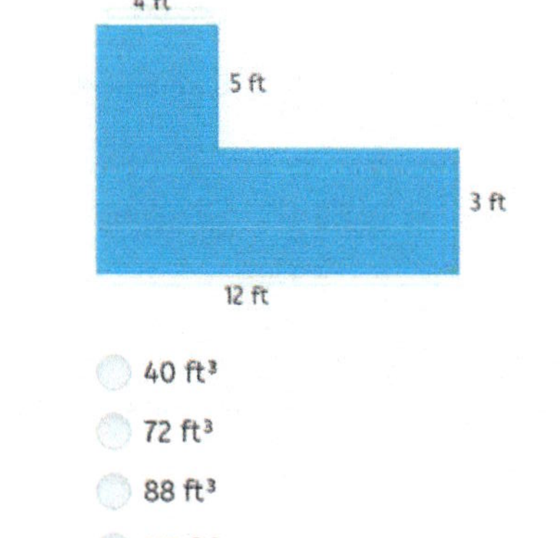

- 40 ft^3
- 72 ft^3
- 88 ft^3
- 112 ft^3

Note taking

1a Do you remember?

A hexagon has ☐ angles and ☐ sides.

Note taking

1b Do you remember?

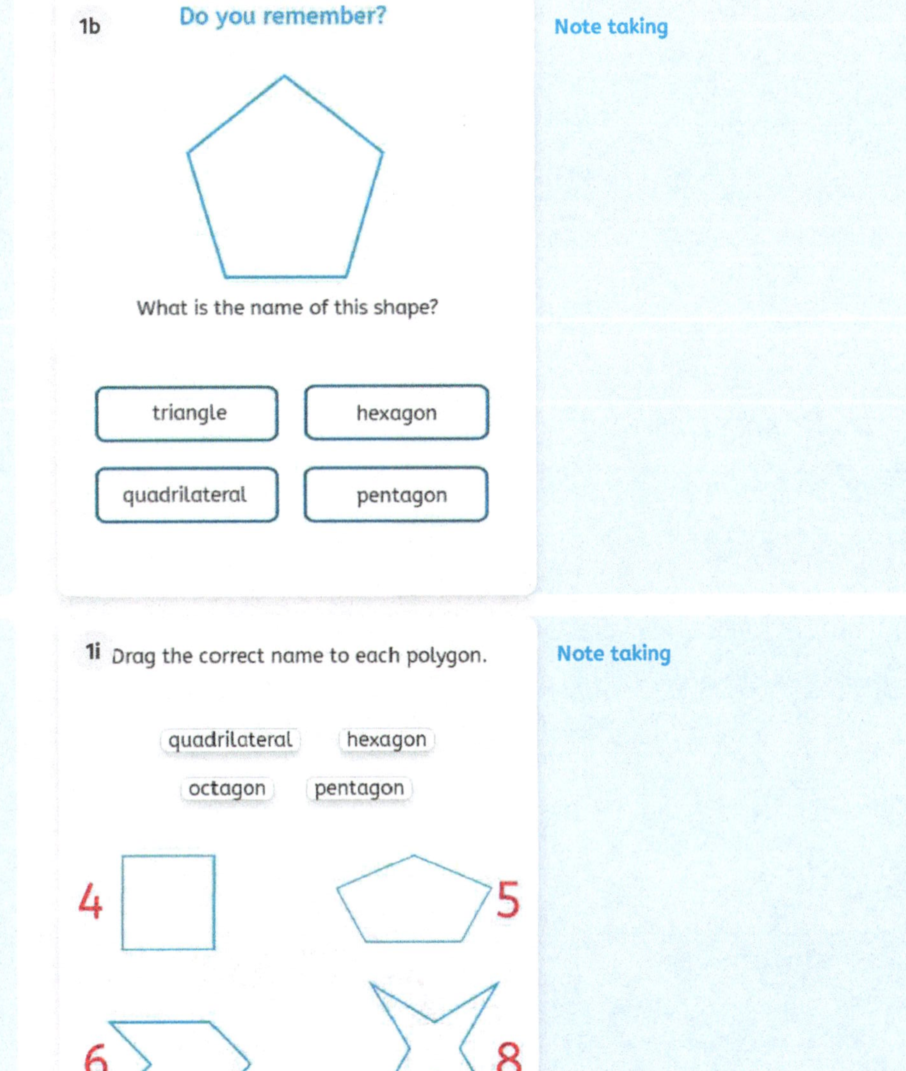

What is the name of this shape?

triangle | hexagon | quadrilateral | pentagon

Note taking

1c Directions

Shape Sort

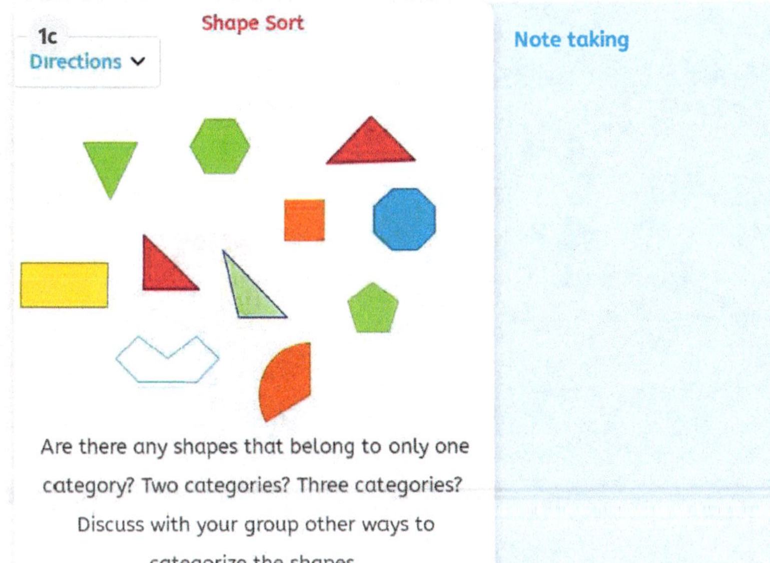

Are there any shapes that belong to only one category? Two categories? Three categories? Discuss with your group other ways to categorize the shapes.

Note taking

1i Drag the correct name to each polygon.

quadrilateral | hexagon | octagon | pentagon

Note taking

1j Which of these shapes is a regular octagon?

Note taking

1k Which statements describe the rectangle?

- No sides are the same length.
- Opposite sides are the same length.
- All sides are the same length.

- No angles are are equal.
- Some angles are equal.
- All angles are equal.

- It is a regular polygon.
- It is not a regular polygon.

Note taking

1l Choose the categories that can be used to classify this figure.

octagon | hexagon | polygon | not regular | regular

Note taking

1m Tap the names that can be used to classify this figure.

triangle | heptagon | quadrilateral | octagon | pentagon | regular | hexagon | not regular

Note taking

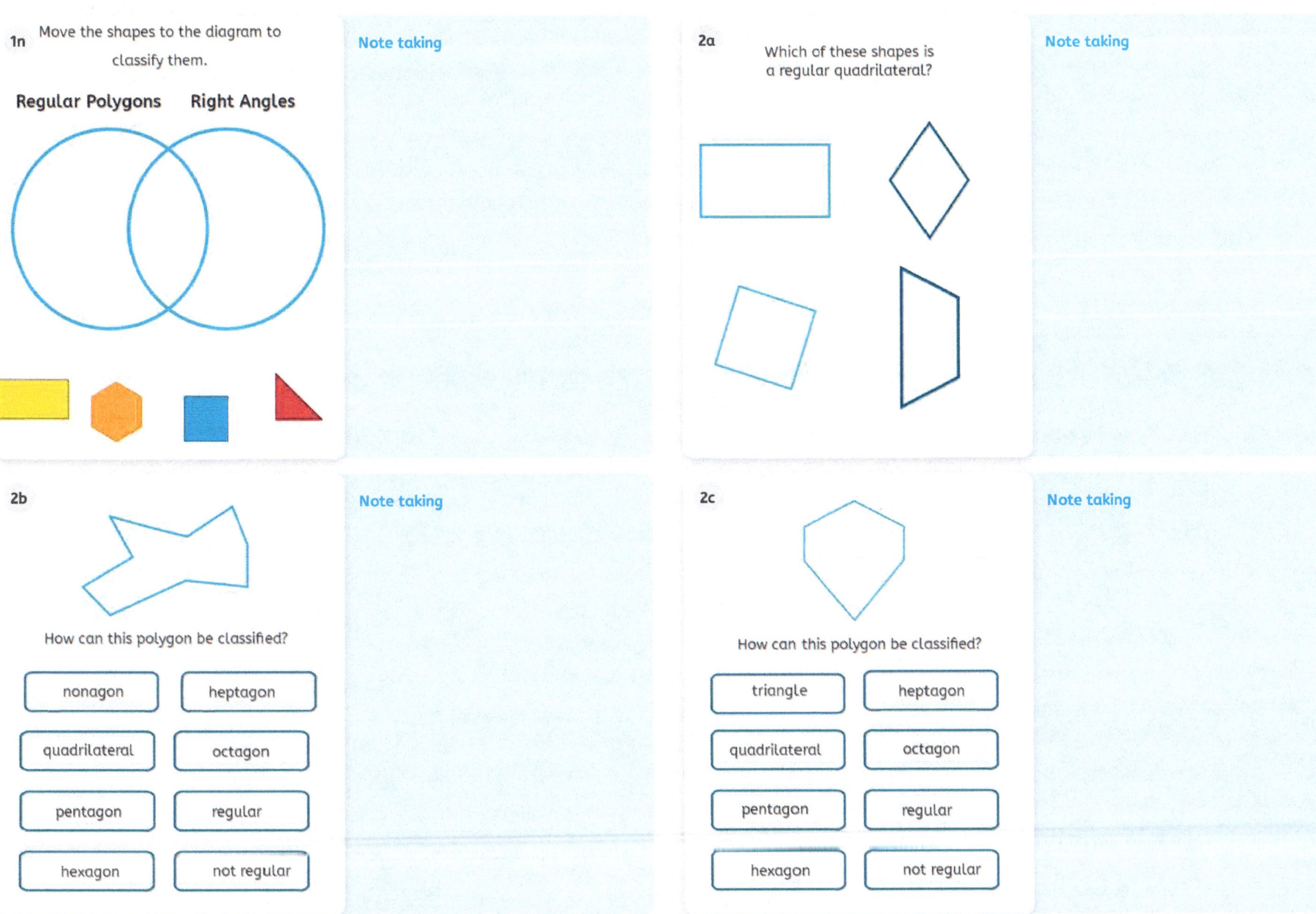
1n
Move the shapes to the diagram to classify them.
Regular Polygons
Right Angles
Note taking
2a
Which of these shapes is a regular quadrilateral?
Note taking
2b
How can this polygon be classified?
nonagon
heptagon
quadrilateral
octagon
pentagon
regular
hexagon
not regular
Note taking
2c
How can this polygon be classified?
triangle
heptagon
quadrilateral
octagon
pentagon
regular
hexagon
not regular
Note taking

2d

How can this polygon be classified?

- triangle
- heptagon
- quadrilateral
- octagon
- pentagon
- regular
- hexagon
- not regular

Note taking

2e

How can this polygon be classified?

- triangle
- heptagon
- quadrilateral
- octagon
- pentagon
- regular
- nonagon
- not regular

Note taking

2f Choose the categories that can be used to classify this figure.

- no sides are the same length
- some sides are the same length
- all sides are the same length

- no angles are equal
- some angles are equal
- all angles are equal

- a regular polygon
- not a regular polygon

Note taking

2g Choose the categories that can be used to classify this figure.

- no sides are the same length
- some sides are the same length
- all sides are the same length

- no angles are equal
- some angles are equal
- all angles are equal

- a regular polygon
- not a regular polygon

Note taking

2h Which of these shapes is a regular hexagon?

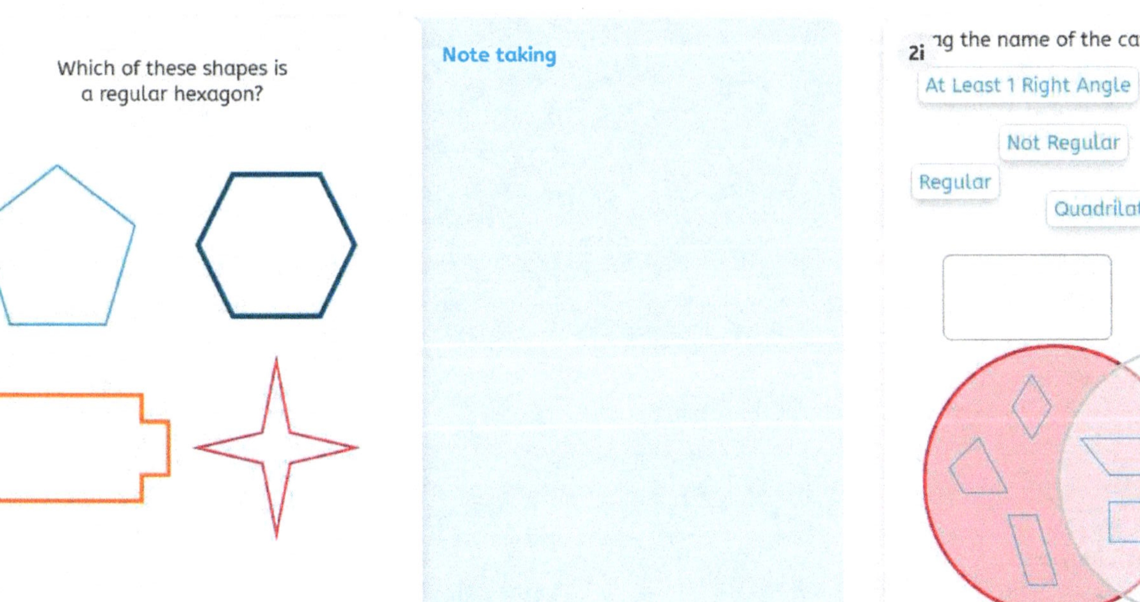

Note taking

2i ...ng the name of the category for each circle

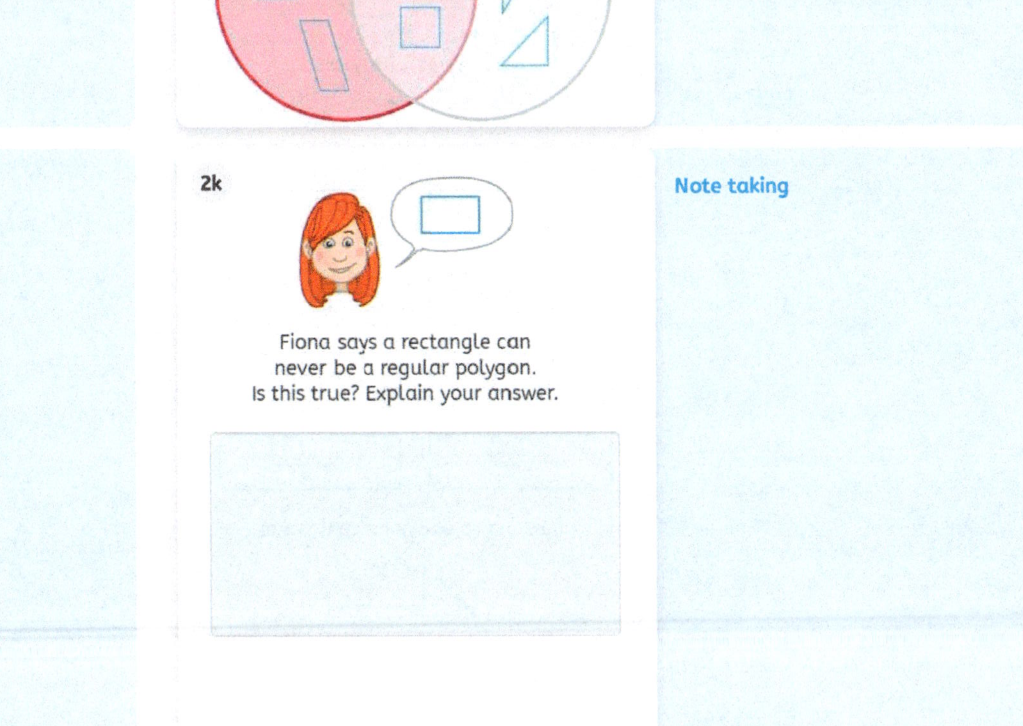

Note taking

2j Move the shapes to the diagram to classify them.

Regular Polygons

At Least 1 Pair of Parallel Sides

Note taking

2k Fiona says a rectangle can never be a regular polygon. Is this true? Explain your answer.

Note taking

2l

The window of a church has the shape of a polygon. It has seven sides and seven angles. The sides are all the same length and the angles have the same measurement.

What is the best name for the shape of the church window?

- A regular hexagon
- An irregular hexagon
- A regular heptagon
- An irregular heptagon

Note taking

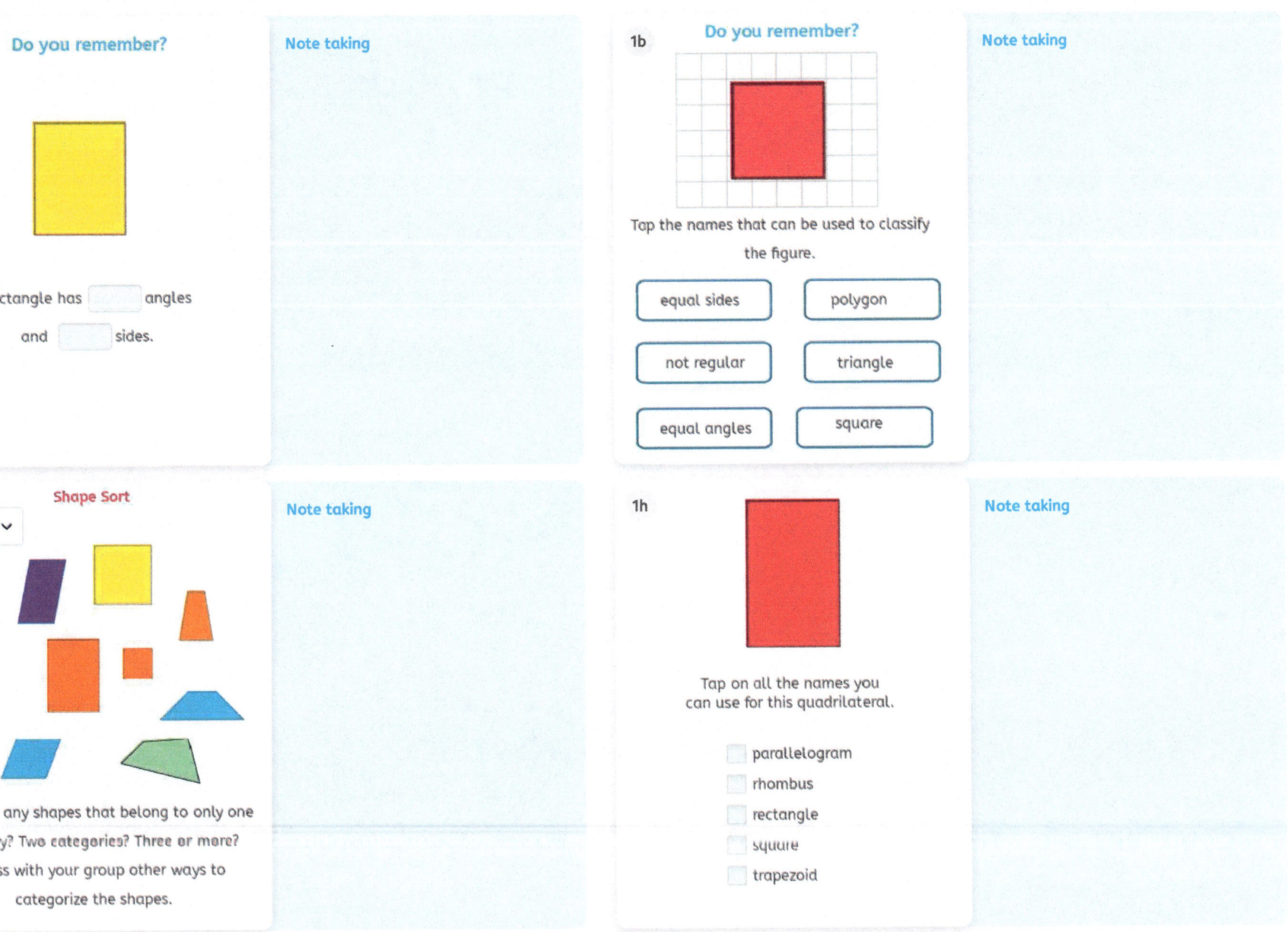
1a
Do you remember?
A rectangle has angles
and sides.
Note taking
1b
Do you remember?
Tap the names that can be used to classify the figure.
equal sides
polygon
not regular
triangle
equal angles
square
Note taking
1c
Directions
Shape Sort
Are there any shapes that belong to only one category? Two categories? Three or more? Discuss with your group other ways to categorize the shapes.
Note taking
1h
Tap on all the names you can use for this quadrilateral.
parallelogram
rhombus
rectangle
square
trapezoid
Note taking

1i

What is the name of this quadrilateral?

- parallelogram
- rhombus
- rectangle
- square
- trapezoid

Note taking

1j What shapes can be called a rhombus? Tap on the correct quadrilaterals.

Note taking

1k Drag each word to classify the group of shapes.

square
quadrilateral
parallelogram
trapezoid
rhombus
rectangle

Note taking

2a

Tap on all the names you can use for this quadrilateral.

- parallelogram
- rhombus
- rectangle
- square
- trapezoid

Note taking

2b

Tap on all the names you can use for this quadrilateral.

- [ ] parallelogram
- [ ] rhombus
- [ ] rectangle
- [ ] square
- [ ] trapezoid

Note taking

2c

Tap on all the names you can use for this quadrilateral.

- ○ parallelogram
- ○ rhombus
- ○ rectangle
- ○ square
- ○ trapezoid

Note taking

2d

Tap on the name you can use for this quadrilateral.

- ○ parallelogram
- ○ rhombus
- ○ rectangle
- ○ square
- ○ trapezoid

Note taking

2e

Tap on all the names you can use for this quadrilateral.

- [ ] parallelogram
- [ ] rhombus
- [ ] rectangle
- [ ] square
- [ ] trapezoid

Note taking

2f

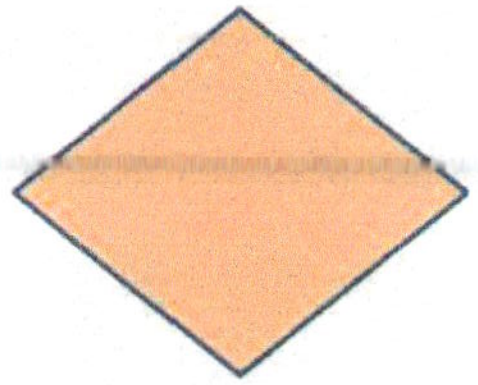

Tap on all the names you can use for this quadrilateral.

- ☐ parallelogram
- ☐ rhombus
- ☐ rectangle
- ☐ square
- ☐ trapezoid

Note taking

2g

Tap on all the names you can use for this quadrilateral.

- ☐ parallelogram
- ☐ rhombus
- ☐ rectangle
- ☐ square
- ☐ trapezoid

Note taking

2h

"I'm thinking about a quadrilateral with just one pair of parallel sides."

What quadrilateral could this be?

- ○ parallelogram
- ○ rhombus
- ○ rectangle
- ○ square
- ○ trapezoid

Note taking

2i

"I'm thinking about a quadrilateral with two pairs of parallel sides and no right angles. Opposite sides are the same length."

What 2 quadrilaterals could this be?

- ☐ parallelogram
- ☐ rhombus
- ☐ rectangle
- ☐ square
- ☐ trapezoid

Note taking

2j

"I'm thinking about a quadrilateral with two pairs of parallel sides. The sides are all the same length."

What quadrilaterals could this be?

- ☐ parallelogram
- ☐ rhombus
- ☐ rectangle
- ☐ square
- ☐ trapezoid

Note taking

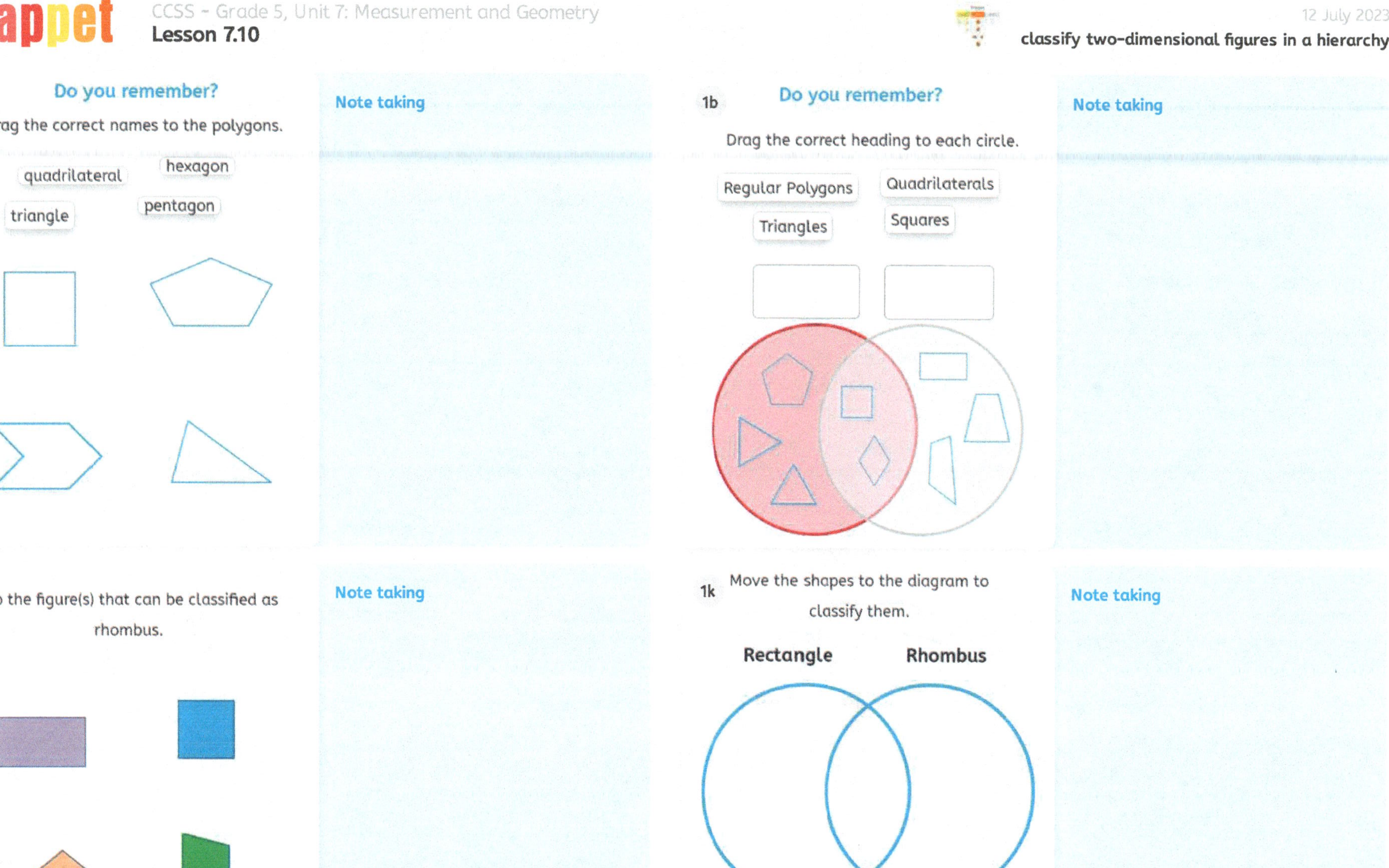

1a Do you remember?

Drag the correct names to the polygons.

quadrilateral, hexagon, triangle, pentagon

Note taking

1b Do you remember?

Drag the correct heading to each circle.

Regular Polygons, Quadrilaterals, Triangles, Squares

Note taking

1j ap the figure(s) that can be classified as rhombus.

Note taking

1k Move the shapes to the diagram to classify them.

Rectangle **Rhombus**

Note taking

1l Drag to shapes to where they belong.

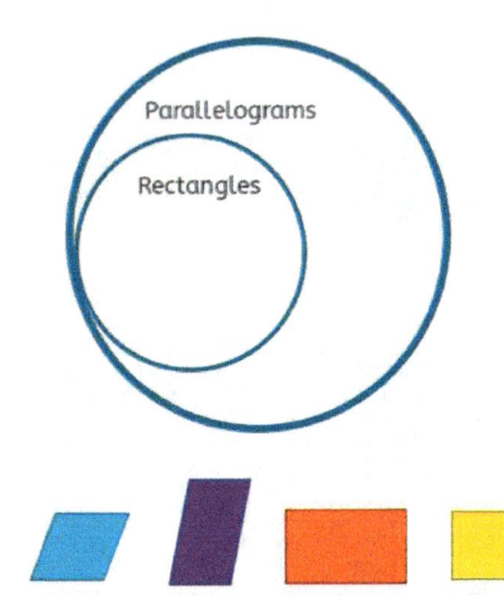

Note taking

1m Drag each label to where it belongs.

Equilateral Triangles

Polygons

Isosceles Triangles

Note taking

1n .iich is the most specific description for <u>all</u> equilateral triangles?

- a 3-sided polygon
- an isosceles triangle with 3 equal sides
- a triangle with a right angle

Note taking

1o Is the statement *always true*, *sometimes true*, or *never true*?

An equilateral triangle is an isosceles triangle.

- always true
- sometimes true
- never true

Note taking

1p Is the statement *always true*, *sometimes true*, or *never true*?

An octagon is a trapezoid.

- always true
- sometimes true
- never true

Note taking

2a ·up the figure(s) that can be classified as both a parallelogram and a square.

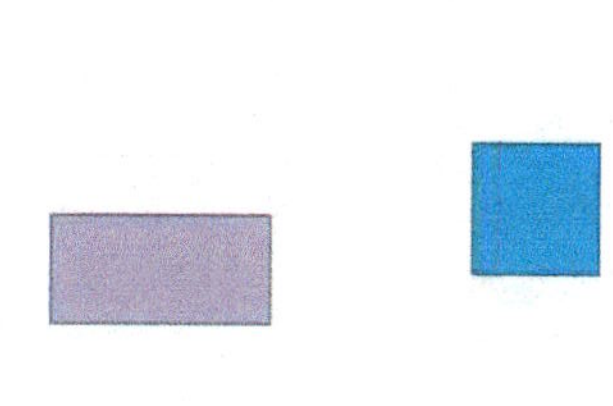

Note taking

2b ·can name a shape from the most general name to the most specific name.

Put a letter on each shape to show where it goes in the diagram.

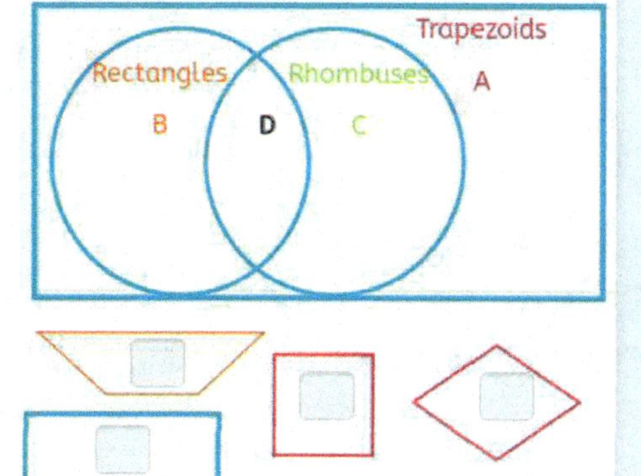

Note taking

2c Which shape is in the wrong part of the diagram?

Tip

Right Triangles

Isosceles Triangles

A

B

C

D

- A
- B
- C
- D

Note taking

2d Drag the correct name to label each circle.

Scalene Triangles | Equilateral Triangles | Isosceles Triangles

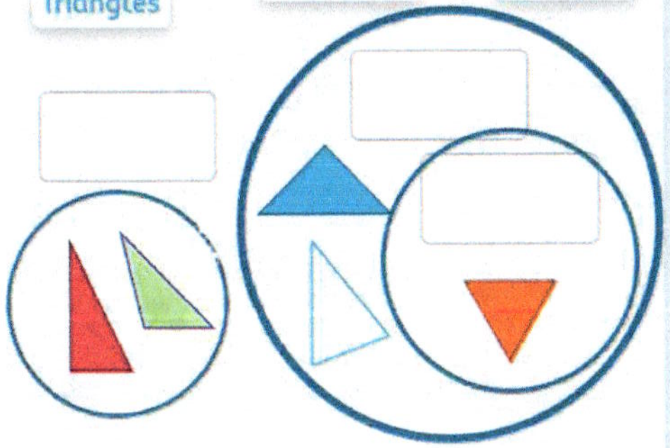

Note taking

2e Drag the names to the diagram.

Polygons → Triangles, Quadrilaterals
Triangles → [ ], Scalene
[ ] → Equilateral
Quadrilaterals → Trapezoids, [ ]
[ ] → Rectangles, Rhombuses
Rectangles, Rhombuses → [ ]

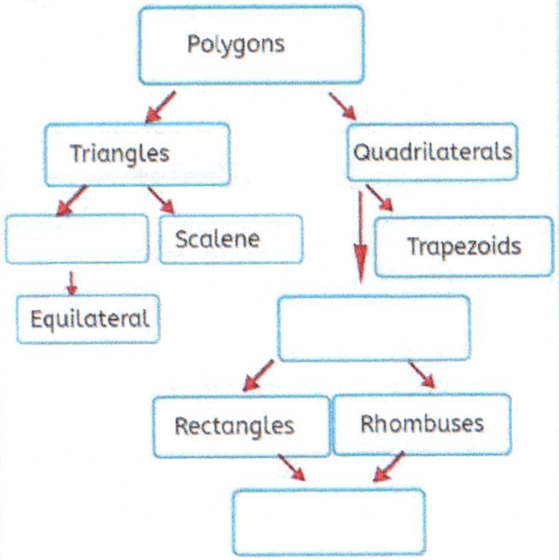

Isosceles | Parallelograms | Squares

Note taking

2f Drag the names to the diagram to show how a rhombus can be classified.

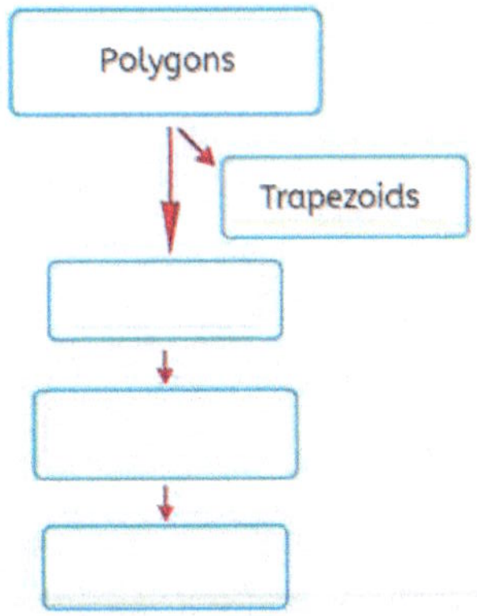

Polygons → Trapezoids, [ ]
[ ] → [ ] → [ ]

Rhombuses | Quadrilaterals | Parallelograms

Note taking

2g Drag the names to the diagram to show how a square can be classified.

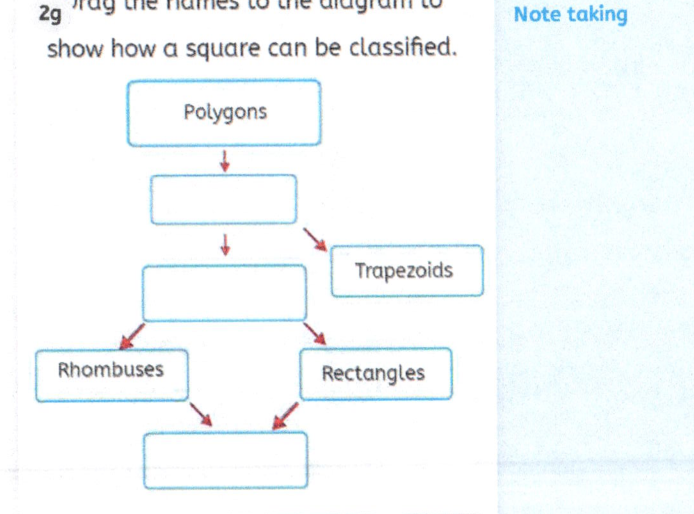

Quadrilaterals | Parallelograms | Squares

Note taking

2h Drag the names to the diagram.

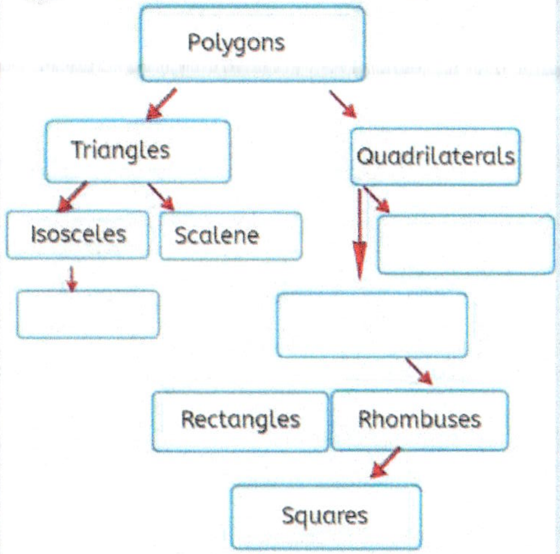

Polygons
Triangles
Quadrilaterals
Isosceles
Scalene
Rectangles
Rhombuses
Squares

Equilateral Parallelograms Trapezoids

Note taking

2i ch is the most specific description for <u>all</u> pentagons?

- a 5-sided polygon
- a closed shape with 1 pair of parallel sides
- a closed shape with 2 right angles

Note taking

2j Is the statement *always true*, *sometimes true*, or *never true*?

A rhombus is a parallelogram.

- always true
- sometimes true
- never true

Note taking

2k Is the statement *always true*, *sometimes true*, or *never true*?

An isosceles triangle is an equilateral triangle.

- always true
- sometimes true
- never true

Note taking

2L Is the statement *always true*, *sometimes true*, or *never true*?

A rectangle is a square.

- always true
- sometimes true
- never true

Note taking

1a

Do you remember?

How many students read three books?

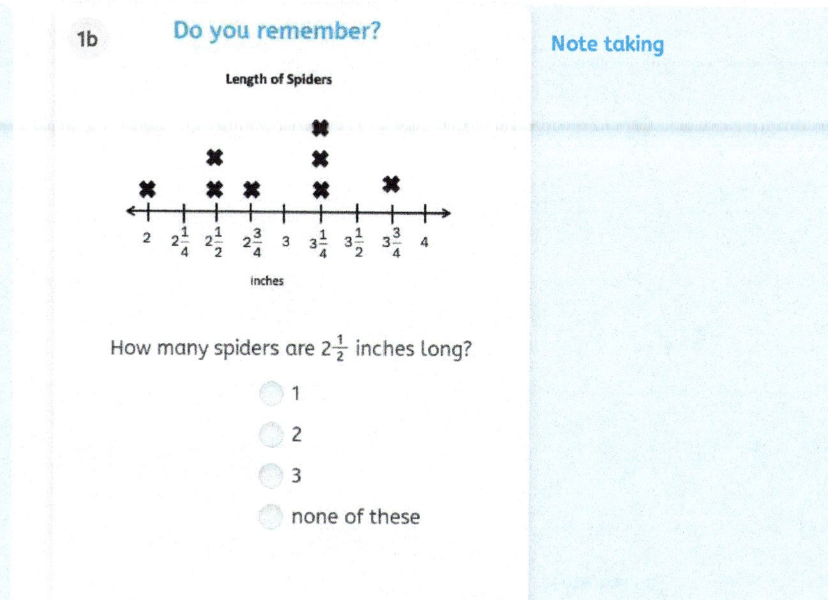

- 0
- 1
- 2
- 3

Note taking

1b

Do you remember?

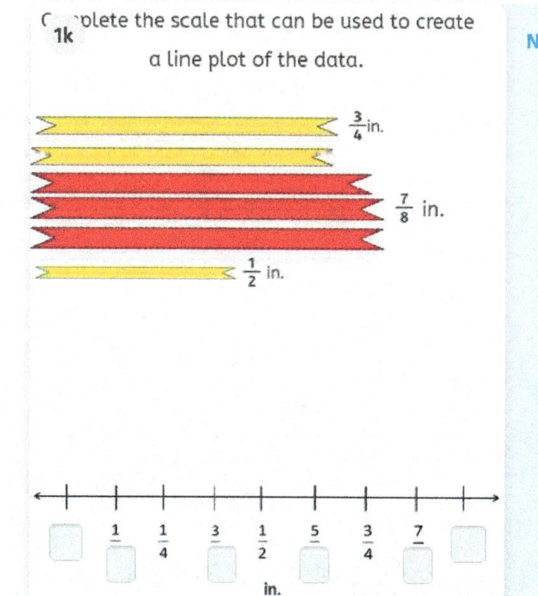

How many spiders are $2\frac{1}{2}$ inches long?

- 1
- 2
- 3
- none of these

Note taking

1j

Ti... cuts tiny pieces of ribbon to decorate her book

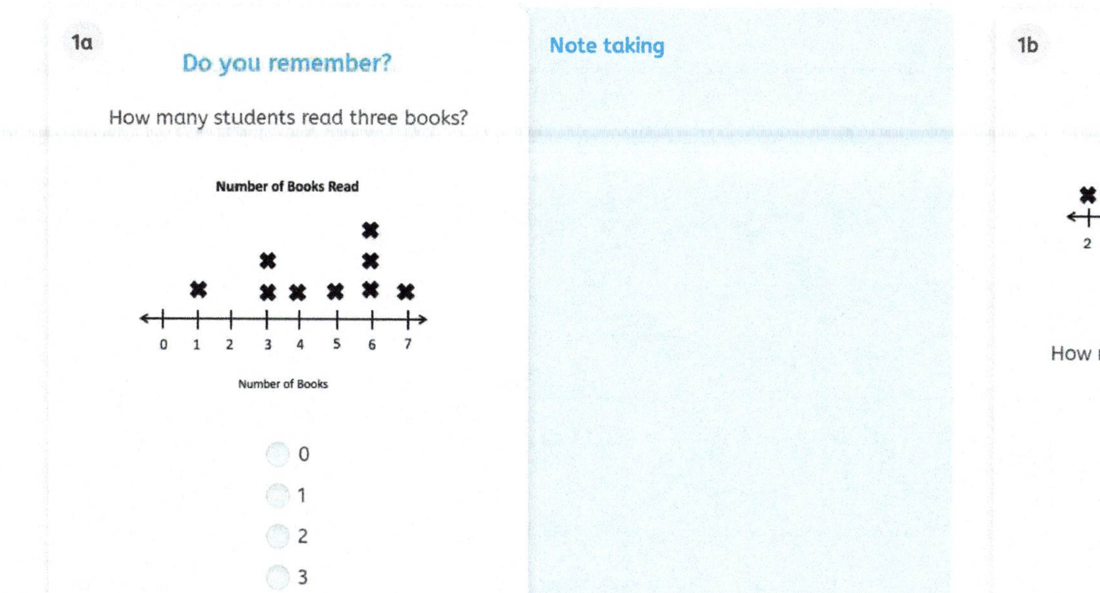

What is the length of the shortest ribbon? $\frac{1}{\square}$ in.

What is the length of the longest ribbon? $\frac{7}{\square}$ in.

What is an appropriate scale for a line plot of the data? $\frac{1}{\square}$ in.

Note taking

1k

C...plete the scale that can be used to create a line plot of the data.

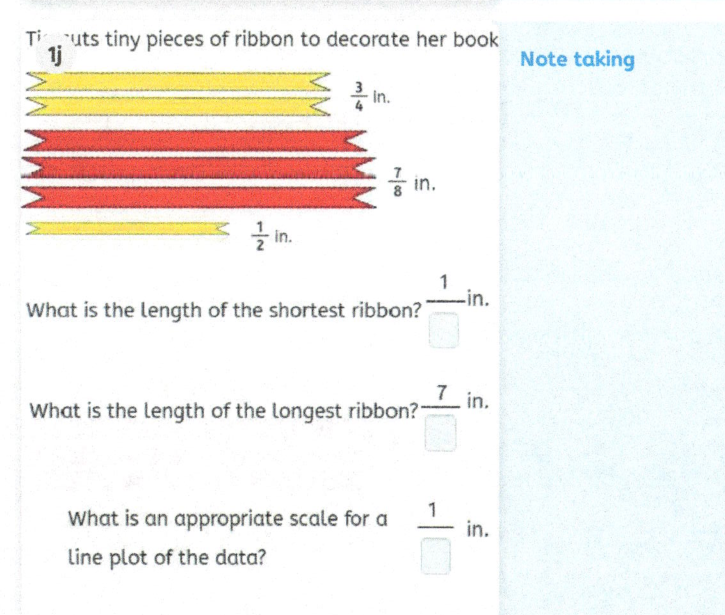

Note taking

1l Create a line plot of the data.

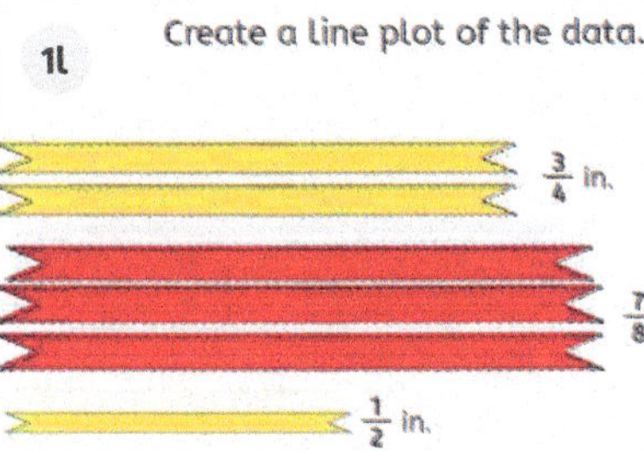

Put an x in the correct boxes to create the line plot.

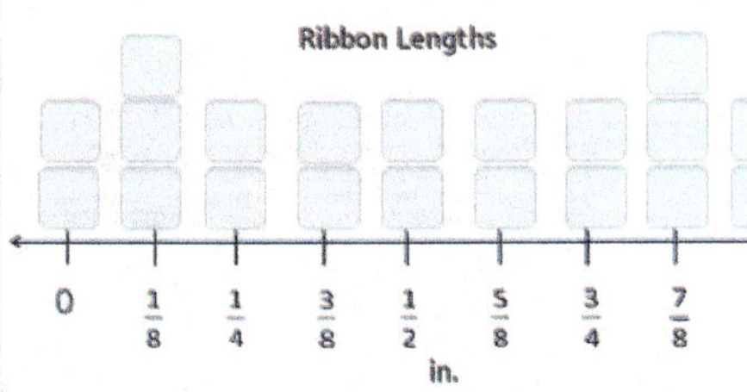

Note taking

1m 5, 5, 5, $5\frac{1}{2}$, $5\frac{1}{2}$, $5\frac{1}{2}$, $5\frac{1}{2}$, $5\frac{1}{2}$

$6\frac{1}{2}$, $6\frac{1}{2}$, 7, 7

Drag the values to answer the questions.

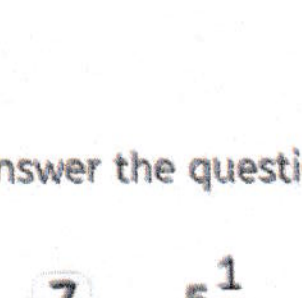
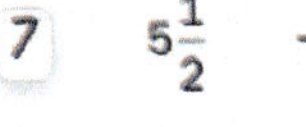

6 | 5 | $6\frac{1}{2}$ | 7 | $5\frac{1}{2}$ | $\frac{1}{2}$

What is the least data value?

What is the greatest data value?

What is an appropriate scale to create a line plot of the data?

Note taking

1n Make a line plot of the data using the X.

5, 5, 5, $5\frac{1}{2}$, $5\frac{1}{2}$, $5\frac{1}{2}$, $5\frac{1}{2}$, $5\frac{1}{2}$

6, 6, $6\frac{1}{2}$, $6\frac{1}{2}$, $6\frac{1}{2}$, 7, 7

✖

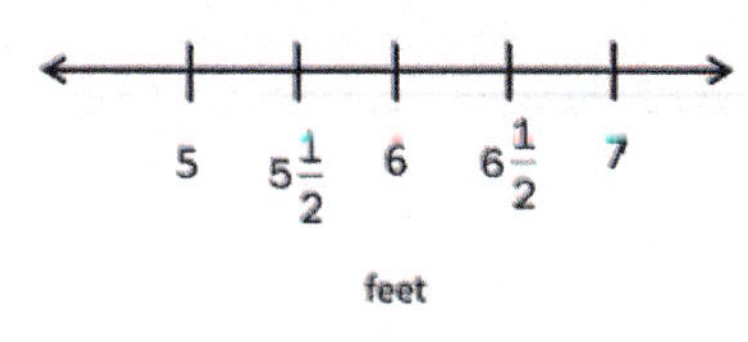

Note taking

2a I... ts tiny pieces of ribbon to use for decoration.

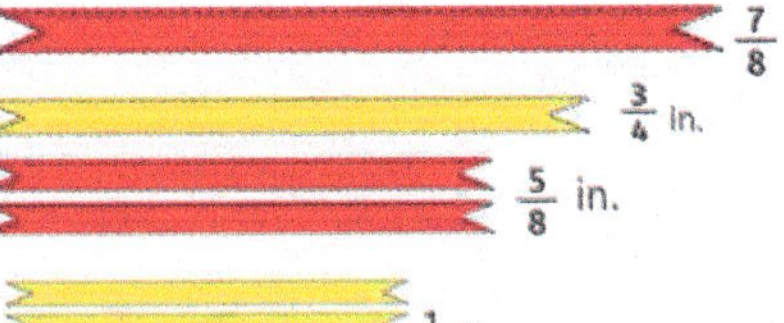

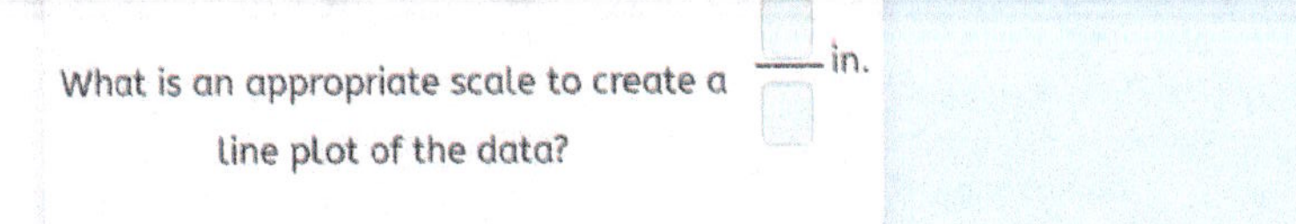

What is the length of the shortest ribbon? ___ in.

What is the length of the longest ribbon? ___ in.

What is an appropriate scale to create a line plot of the data? ___ in.

Note taking

2b C…plete the scale that can be used to create a dot plot of the data.

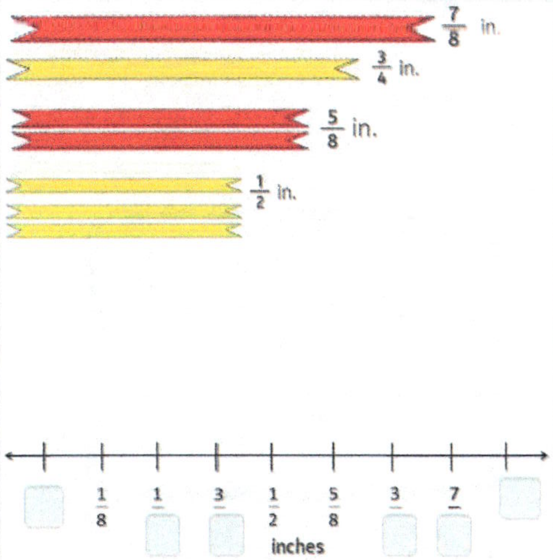

$\frac{7}{8}$ in.
$\frac{3}{4}$ in.
$\frac{5}{8}$ in.
$\frac{1}{2}$ in.

$\frac{1}{8}$ 1 3 $\frac{1}{2}$ $\frac{5}{8}$ 3 7

inches

Note taking

2c Put an x in the correct boxes to create a line plot of the data.

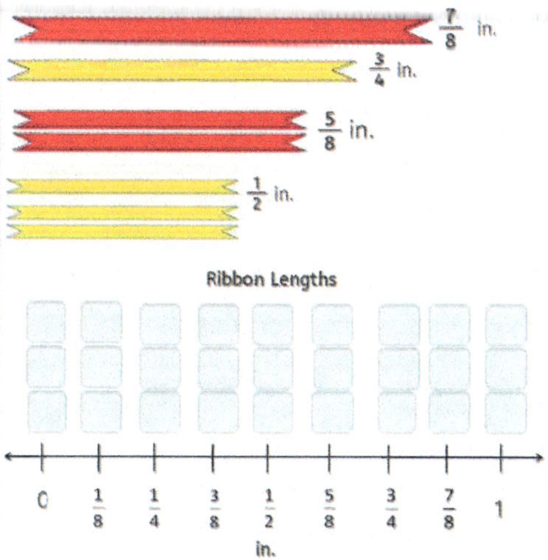

$\frac{7}{8}$ in.
$\frac{3}{4}$ in.
$\frac{5}{8}$ in.
$\frac{1}{2}$ in.

Ribbon Lengths

0 $\frac{1}{8}$ $\frac{1}{4}$ $\frac{3}{8}$ $\frac{1}{2}$ $\frac{5}{8}$ $\frac{3}{4}$ $\frac{7}{8}$ 1

in.

Note taking

2d …ag Xs to the line plot to show the weights.

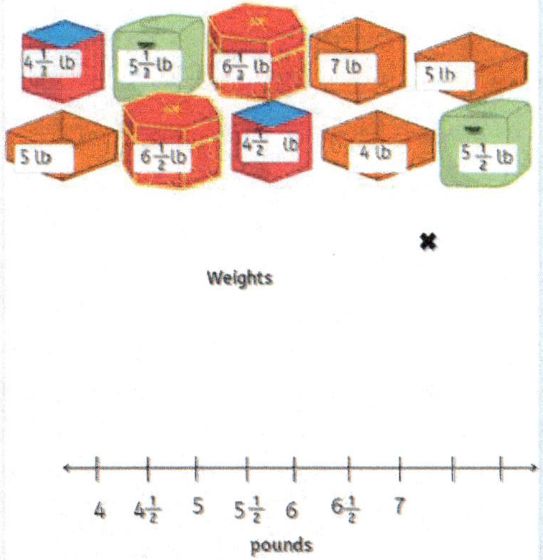

$4\frac{1}{2}$ lb, $5\frac{1}{2}$ lb, $6\frac{1}{2}$ lb, 7 lb, 5 lb, 5 lb, $6\frac{1}{2}$ lb, $4\frac{1}{2}$ lb, 4 lb, $5\frac{1}{2}$ lb

Weights

4 $4\frac{1}{2}$ 5 $5\frac{1}{2}$ 6 $6\frac{1}{2}$ 7

pounds

Note taking

2e Use the data in the table to make a line plot.

Flower length	Tally
4 inches	|||
5 inches	||
6 inches	||||
7 inches	|||

drag

X

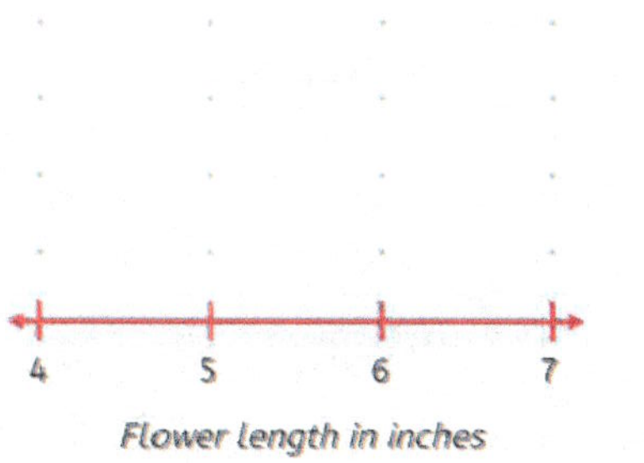

4 5 6 7

Flower length in inches

Note taking

2f ᵉ the data in the table to make a line plot.

Ribbon Length (in.)	Number of Ribbons
$5\frac{1}{2}$	2
$6\frac{1}{2}$	4
$7\frac{1}{2}$	3
$8\frac{1}{2}$	1

drag

X

$5\frac{1}{2}$ $6\frac{1}{2}$ $7\frac{1}{2}$ $8\frac{1}{2}$

Ribbon length in inches

Note taking

2g ᵉ the data in the table to make a line plot.

Bookshelf Length (yd)	Number of Shelves
$\frac{1}{8}$	4
$\frac{3}{8}$	2
$\frac{5}{8}$	1
$\frac{7}{8}$	4

drag

X

0 $\frac{1}{8}$ $\frac{1}{4}$ $\frac{3}{8}$ $\frac{1}{2}$ $\frac{5}{8}$ $\frac{3}{4}$ $\frac{7}{8}$ 1

Length of bookshelves

Note taking

2h 'ace an **x** in the boxes to create a line plot of the data.

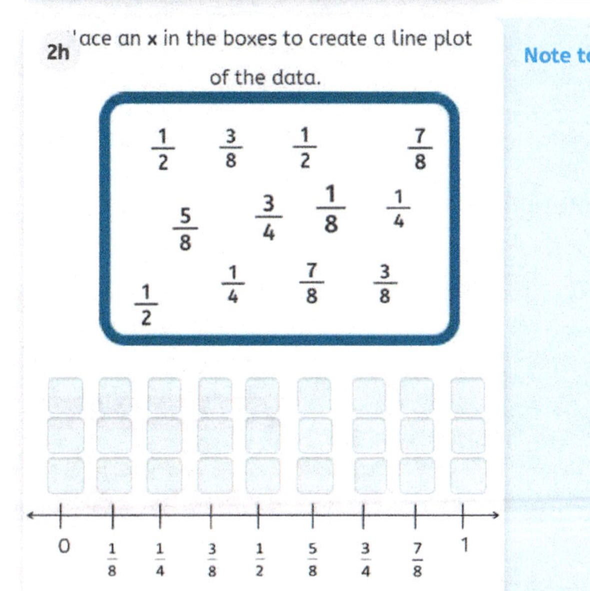

Note taking

1a

Do you remember?

How many beetles are represented in the graph?

Length of Beetles

2, $2\frac{1}{4}$, $2\frac{1}{2}$, $2\frac{3}{4}$, 3, $3\frac{1}{4}$, $3\frac{1}{2}$, $3\frac{3}{4}$, 4

cm

- 7
- 8
- 9
- 10

Note taking

1b

Do you remember?

$\frac{1}{4} + \frac{3}{8} = ?$

$\frac{\square}{8} + \frac{3}{8} = \frac{\square}{8}$

Note taking

1c

Do you remember?

$\frac{3}{4} \times 6 = \frac{\square}{\square}$

Note taking

1d

The line plot shows the lengths of yarn Cam has. What is the difference between the longest and shortest pieces?

Steps

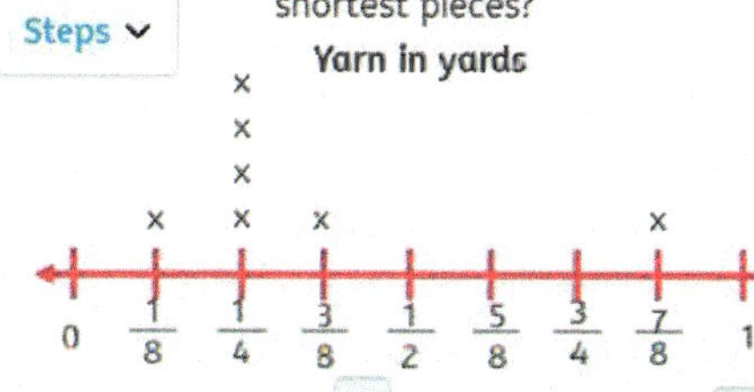

The longest piece is $\frac{\square}{\square}$ yd. The shortest is $\frac{\square}{\square}$ yd.

The difference is $\frac{\square}{8}$ $\square$ $\frac{\square}{\square} = \frac{\square}{\square}$ yd

Note taking

1m

The line plot shows Cam's measuring cups. Cam needs five times the amount of flour that the smallest cup holds.

How much flour does Cam need?

Steps

Measuring Cups, in cups

$\frac{1}{8}$, $\frac{1}{4}$, $\frac{3}{8}$, $\frac{1}{2}$

The smallest cup holds ___ cup.

Cam needs 5 x ___ = ___ cups of flour.

Note taking

1n

Pieces of yarn in yards

0, $\frac{1}{4}$, $\frac{1}{2}$, $\frac{3}{4}$

What is the total of the yarn that are $\frac{1}{4}$ yd long?

___ pieces are $\frac{1}{4}$ yd long.

The pieces have a total length of ___ yard.

Note taking

1o

Length of yarn in yards

0, $\frac{1}{4}$, $\frac{1}{2}$, $\frac{3}{4}$

The $\frac{3}{4}$-yd pieces are joined, and then cut into 3 equal pieces. What is the length of each piece?

___ pieces are $\frac{3}{4}$ yd long.

They have a total length of ___ yd.

___ yd ÷ ___ = ___ yd

Each piece is ___ yd.

Note taking

2a

The line plot shows the lengths of yarn Cam has. What is the difference between the longest and shortest pieces?

Steps

Yarn in yards

0, $\frac{1}{8}$, $\frac{1}{4}$, $\frac{3}{8}$, $\frac{1}{2}$, $\frac{5}{8}$, $\frac{3}{4}$, $\frac{7}{8}$, 1

The longest piece is $\frac{\square}{4}$ yd. The shortest is ___ yd.

The difference is: ___ - ___ = ___ yd

Note taking

2b

The line plot shows the lengths of yarn Cam has. What is the difference between the longest and shortest pieces?

Steps

Yarn in yards

0, $\frac{1}{8}$, $\frac{1}{4}$, $\frac{3}{8}$, $\frac{1}{2}$, $\frac{5}{8}$, $\frac{3}{4}$, $\frac{7}{8}$, 1

The longest piece is ___ yd. The shortest is ___ yd.

The difference is: ___ ___ ___ = ___ yd.

Note taking

2c

The line plot shows Cam's measuring cups. Cam needs three times the amount of flour that the smallest cup holds.

How much flour does Cam need?

Steps

Measuring Cups, in cups

$\frac{1}{4}$, $\frac{1}{2}$, $\frac{3}{4}$, 1

The smallest cup holds ___ cup.

Cam needs 3 x ___ = ___ cups of flour.

Note taking

2d

The line plot shows Cam's measuring cups. Cam needs two times the amount of flour that the smallest cup holds.

How much flour does Cam need?

Steps

Measuring Cups, in cups

$\frac{3}{8}$, $\frac{1}{2}$, $\frac{5}{8}$, $\frac{6}{8}$

The smallest cup holds ___ cup.

Cam needs 2 x ___ = $\frac{\square}{8}$ cups of flour.

Note taking

2e

Pieces of yarn in yards

$\frac{1}{4}$, $\frac{3}{8}$, $\frac{1}{2}$, $\frac{5}{8}$

How many pieces of yarn are $\frac{3}{8}$ yd long?

___ pieces are $\frac{3}{8}$ yd long.

Those pieces have a total length of $\frac{6}{\square}$ yard.

Note taking

2f

Pieces of yarn in yards

Line plot: $\frac{1}{4}$: x x; $\frac{3}{8}$: none; $\frac{1}{2}$: x x x x; $\frac{5}{8}$: x x x

How many pieces of yarn are $\frac{5}{8}$ yd long?

☐ pieces are $\frac{5}{8}$ yd long.

They have a total length of $\frac{☐}{8}$ yard.

Note taking

2g

The line plot shows Cam's measuring cups.
Cam needs three times the amount of flour that the largest cup holds.

How much flour does Cam need?

Measuring Cups, in cups

Line plot: $\frac{3}{8}$: x x; $\frac{1}{2}$: none; $\frac{5}{8}$: x x; $\frac{3}{4}$: x x x x

○ $\frac{9}{4}$
○ $\frac{6}{4}$
○ $\frac{3}{2}$

Note taking

2h

Length of yarn in yards

Line plot: 0; $\frac{1}{8}$: x x x; $\frac{1}{4}$: x; $\frac{3}{8}$: none; $\frac{1}{2}$: x x x x; $\frac{5}{8}$: none; $\frac{3}{4}$: x x; $\frac{7}{8}$: none; 1: x x x

The $\frac{1}{2}$-yd pieces are joined and then cut into four equal pieces. What is the length of each piece?

☐ pieces are $\frac{1}{2}$ yd long.

They have a total length of ☐ yd.

☐ yd ÷ ☐ = $\frac{☐}{☐}$ yd

Each piece is $\frac{☐}{4}$ yd.

Note taking

2i

Length of yarn in yards

Line plot: 0; $\frac{1}{8}$: x; $\frac{1}{4}$: x x x x; $\frac{3}{8}$: x x; $\frac{1}{2}$: none; $\frac{5}{8}$: x x x x; $\frac{3}{4}$: none; $\frac{7}{8}$: x; 1: x x x

The $\frac{1}{4}$-yd pieces are joined and then cut into four equal pieces. What is the length of each piece?

☐ pieces are $\frac{1}{4}$ yd long.

They have a total length of ☐ yd.

☐ yd ÷ ☐ = $\frac{☐}{☐}$ yd

Each piece is $\frac{☐}{☐}$ yd.

Note taking

2j

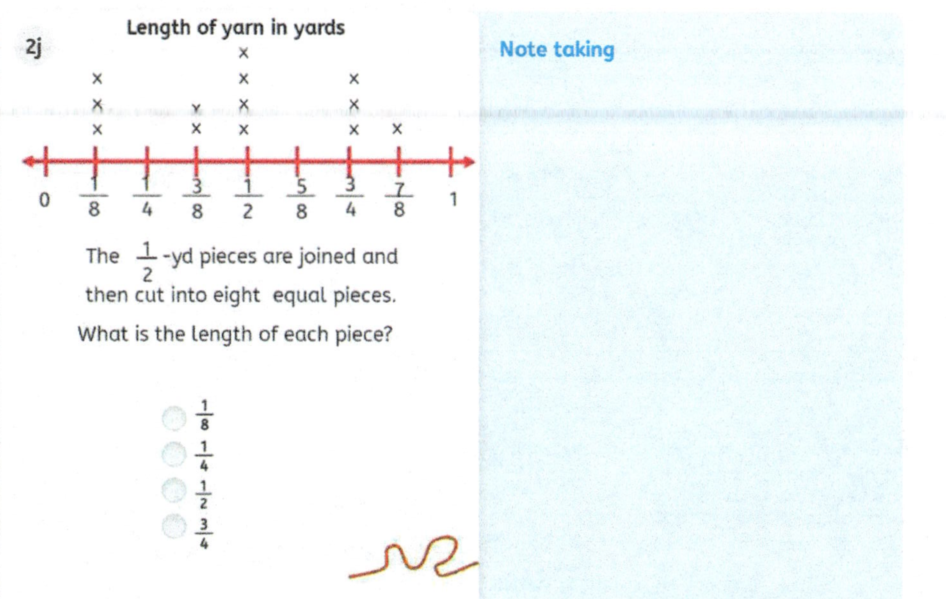

The $\frac{1}{2}$-yd pieces are joined and then cut into eight equal pieces.

What is the length of each piece?

- $\frac{1}{8}$
- $\frac{1}{4}$
- $\frac{1}{2}$
- $\frac{3}{4}$

Note taking

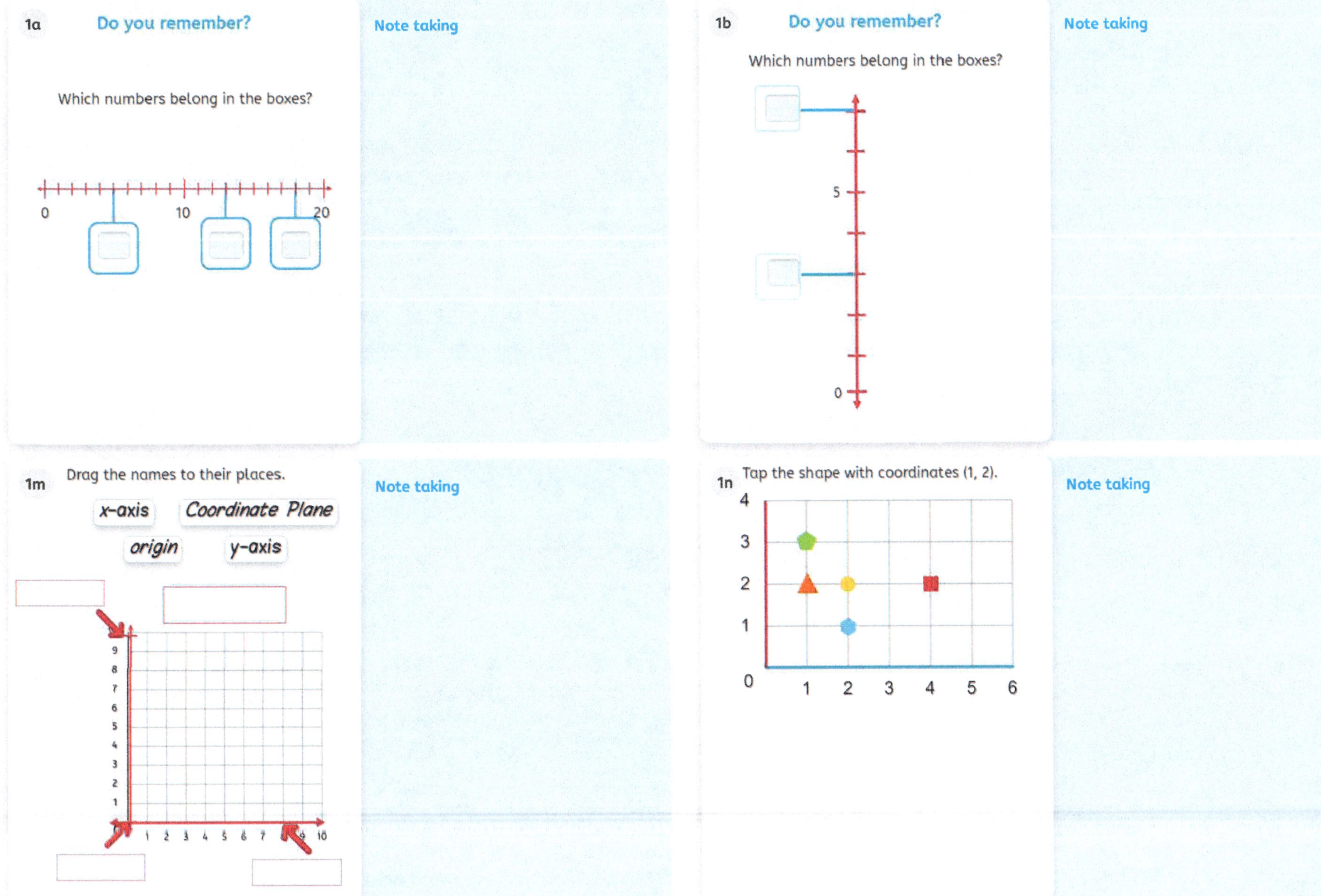
1a
Do you remember?
Which numbers belong in the boxes?
0
10
20
Note taking
1b
Do you remember?
Which numbers belong in the boxes?
5
0
Note taking
1m
Drag the names to their places.
x-axis
Coordinate Plane
origin
y-axis
Note taking
1n
Tap the shape with coordinates (1, 2).
4
3
2
1
0
1
2
3
4
5
6
Note taking

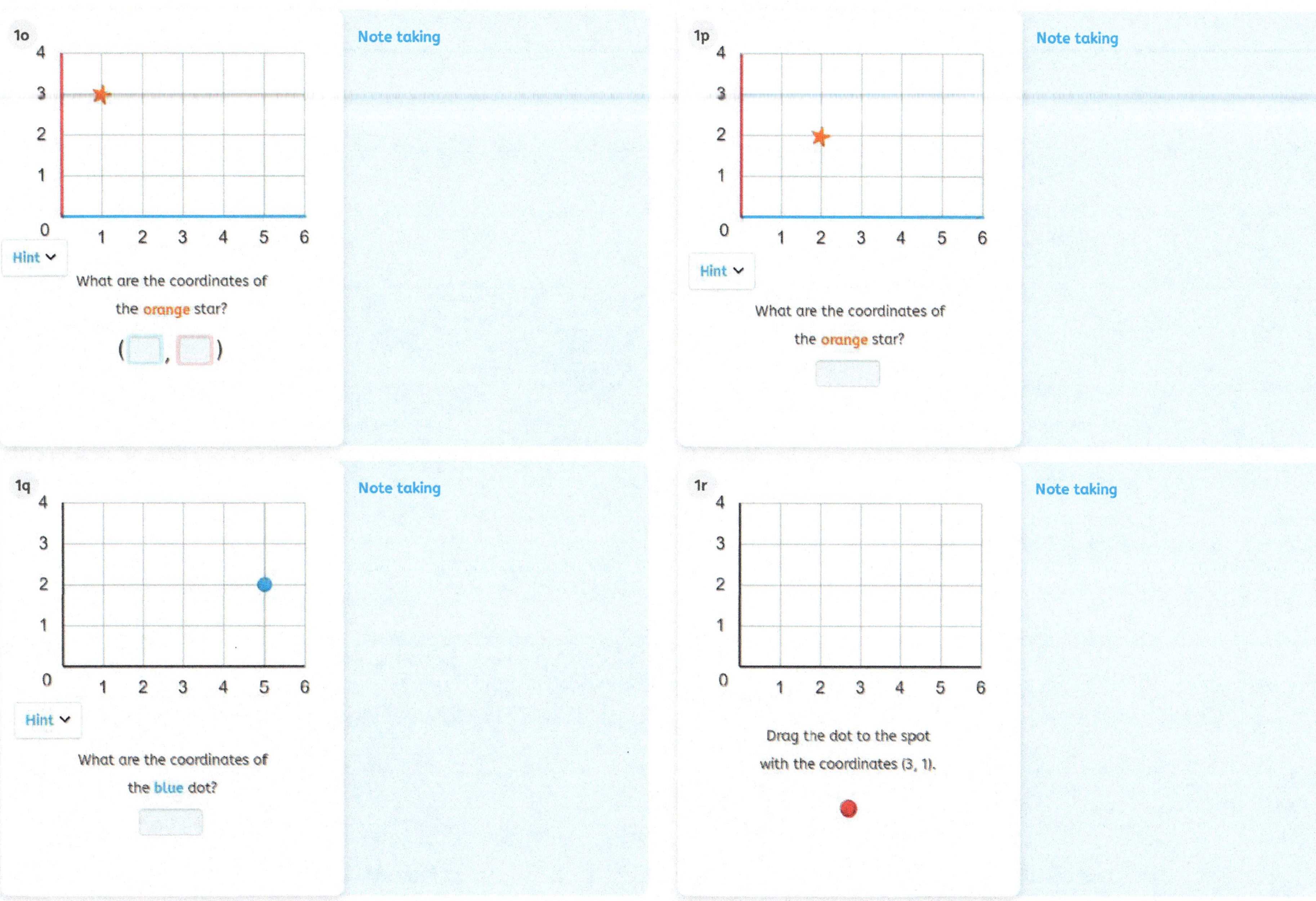
1o
4
3
2
1
0
1 2 3 4 5 6
Hint
What are the coordinates of the orange star?
(,)
Note taking
1p
4
3
2
1
0
1 2 3 4 5 6
Hint
What are the coordinates of the orange star?
Note taking
1q
4
3
2
1
0
1 2 3 4 5 6
Hint
What are the coordinates of the blue dot?
Note taking
1r
4
3
2
1
0
1 2 3 4 5 6
Drag the dot to the spot with the coordinates (3, 1).
Note taking

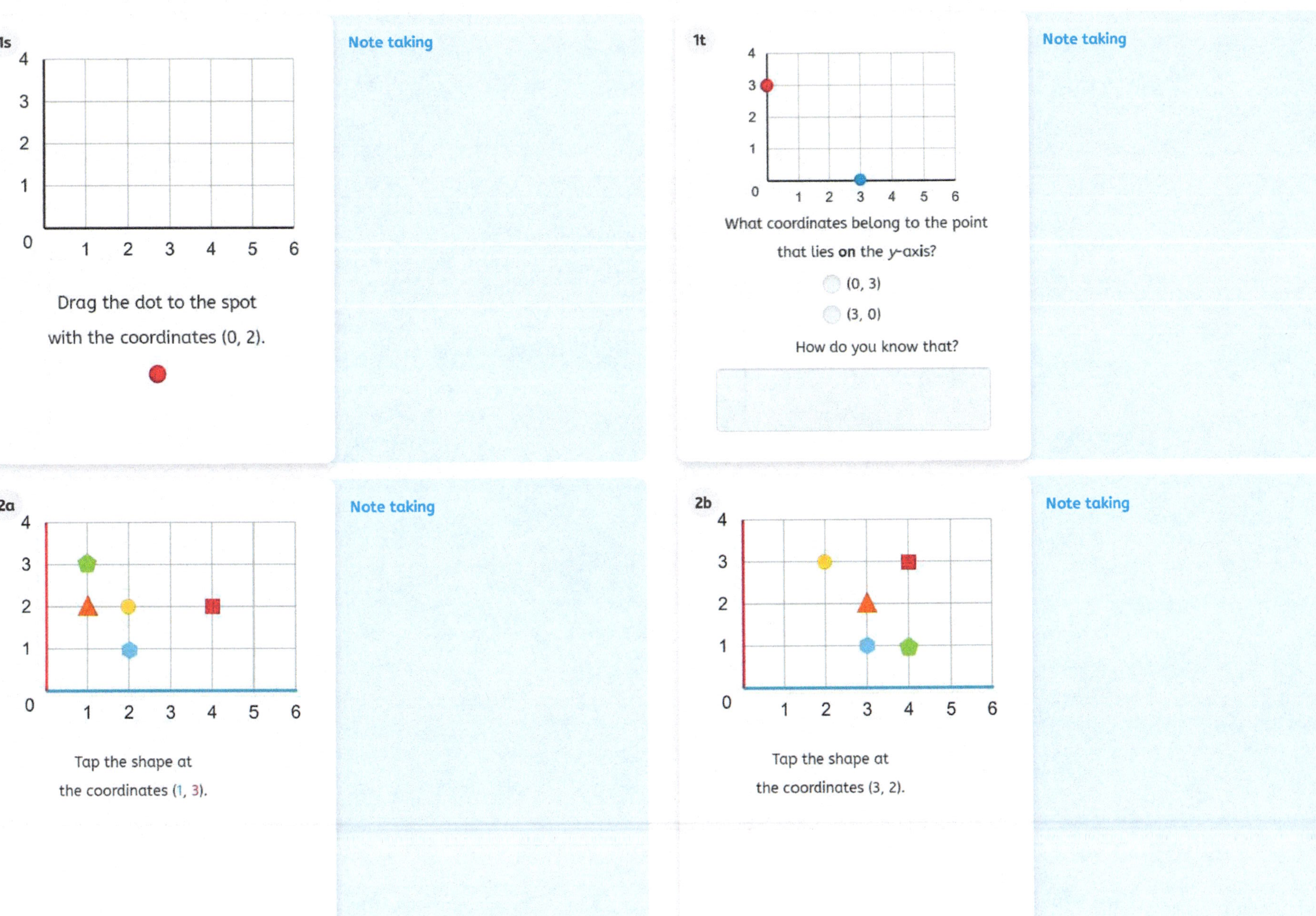
1s
Drag the dot to the spot
with the coordinates (0, 2).
Note taking
1t
What coordinates belong to the point
that lies on the y-axis?
(0, 3)
(3, 0)
How do you know that?
Note taking
2a
Tap the shape at
the coordinates (1, 3).
Note taking
2b
Tap the shape at
the coordinates (3, 2).
Note taking

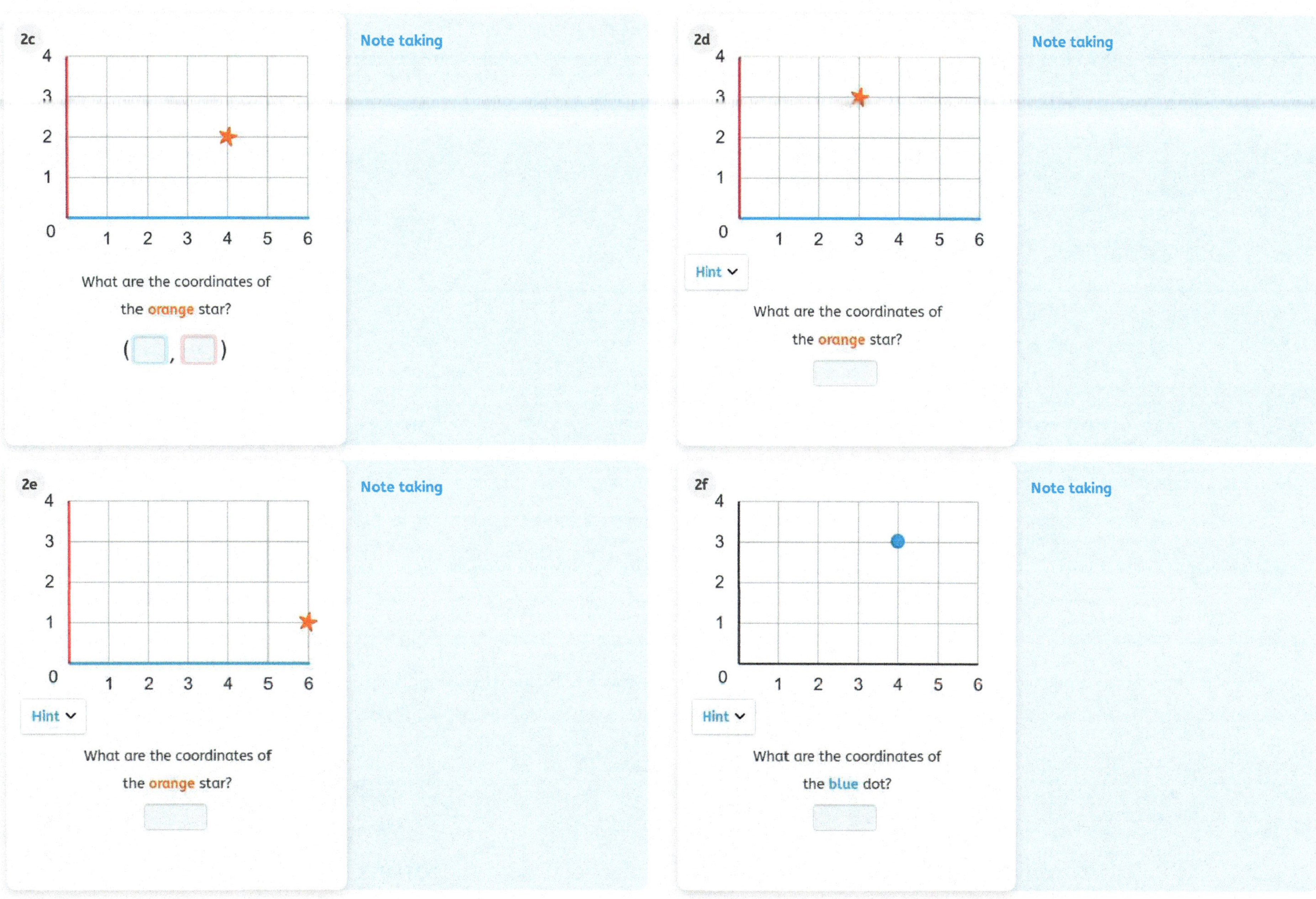
2c
4
3
2
1
0
1 2 3 4 5 6
What are the coordinates of
the orange star?
(,)
Note taking
2d
4
3
2
1
0
1 2 3 4 5 6
Hint
What are the coordinates of
the orange star?
Note taking
2e
4
3
2
1
0
1 2 3 4 5 6
Hint
What are the coordinates of
the orange star?
Note taking
2f
4
3
2
1
0
1 2 3 4 5 6
Hint
What are the coordinates of
the blue dot?
Note taking

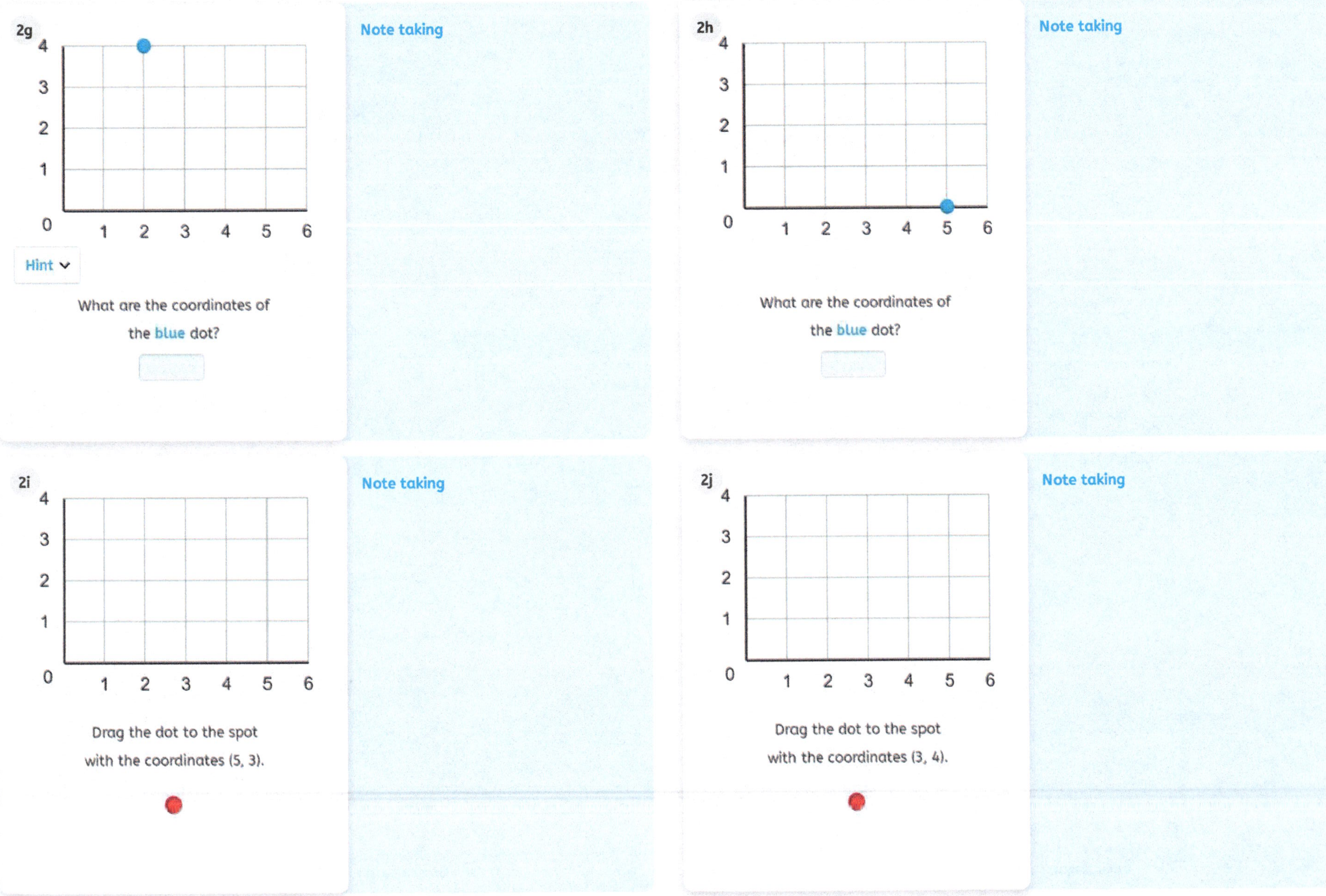
2g
4
3
2
1
0
1 2 3 4 5 6
Hint
What are the coordinates of the blue dot?
Note taking
2h
4
3
2
1
0
1 2 3 4 5 6
What are the coordinates of the blue dot?
Note taking
2i
4
3
2
1
0
1 2 3 4 5 6
Drag the dot to the spot with the coordinates (5, 3).
Note taking
2j
4
3
2
1
0
1 2 3 4 5 6
Drag the dot to the spot with the coordinates (3, 4).
Note taking

2k

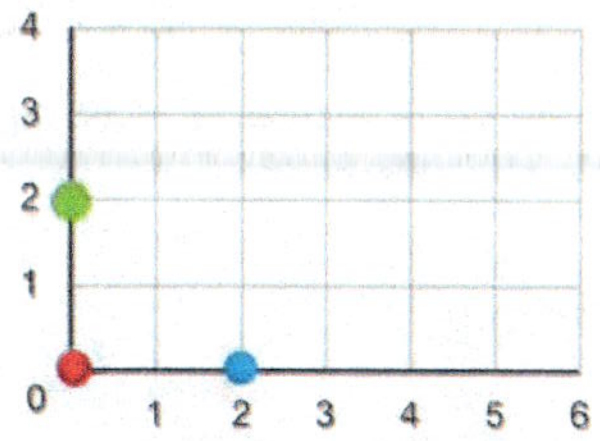

What coordinates belong to the point that lies **at the origin**?

- (0, 2)
- (2, 0)
- (0, 0)

How do you know that?

Note taking

2l

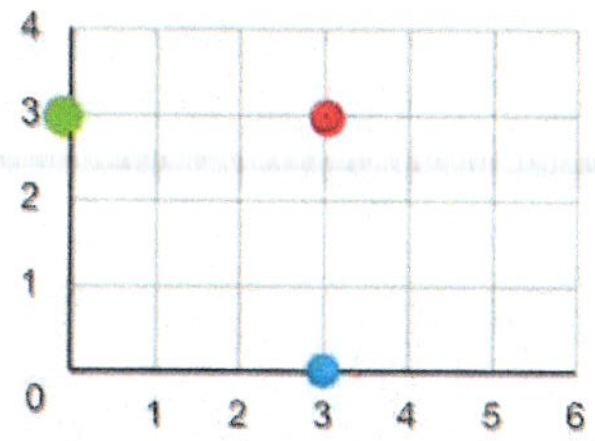

What coordinates belong to the point that lies **on the *x*-axis**?

- (3, 0)
- (0, 3)
- (3, 3)

How do you know that?

Note taking

1a Do you remember?

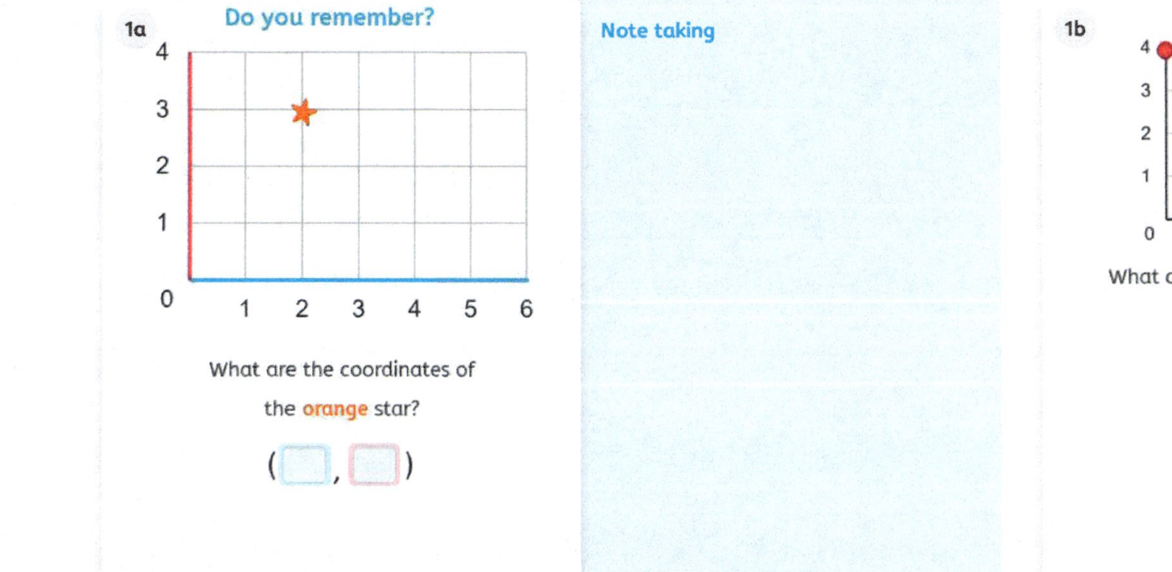

What are the coordinates of the orange star?

(☐ , ☐)

Note taking

1b Do you remember?

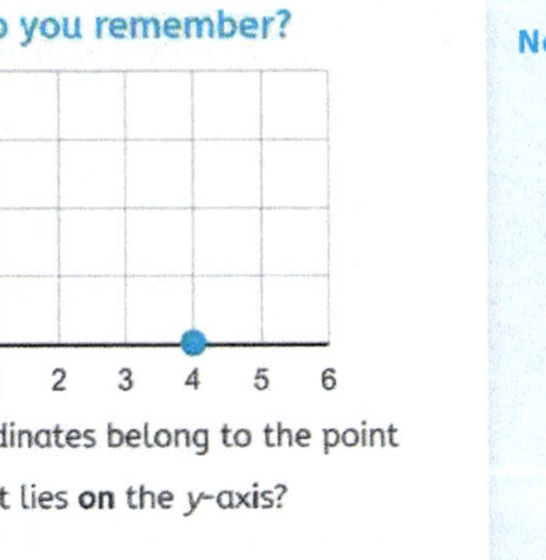

What coordinates belong to the point that lies **on** the *y*-axis?

- (0, 4)
- (4, 0)
- (4, 4)
- (0, 0)

Note taking

1k Ana traveled from the school to the park and then to the library.

How many blocks did she travel?

Steps

blocks
Park
Library
School
blocks

The school to the park is ☐ blocks.

The park to the library is ☐ blocks.

Ana traveled ☐ blocks.

Note taking

1l Ana left the restaurant and went to the gym.

Then she went to the store.

How many blocks did she travel?

Steps

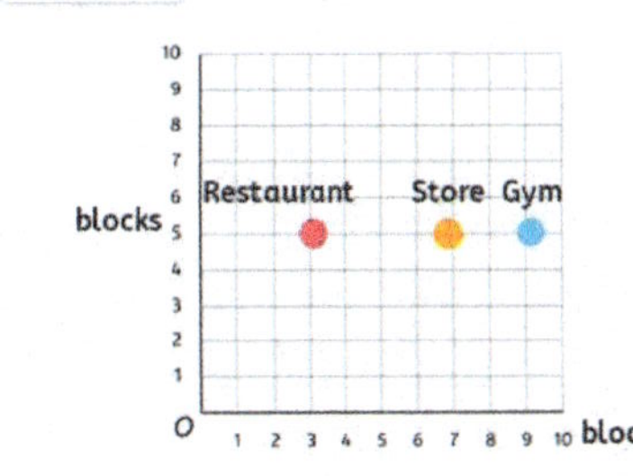

The restaurant to the gym is ☐ blocks.

The gym to the store is ☐ blocks.

Ana traveled ☐ blocks.

Note taking

1m ...m left the school located at (3, 4). He went to the park located at (6, 4). Then he went to the library located at (6, 8).

How many blocks did he travel?

Use the coordinate plane to help you.

From the school to the park is ☐ blocks.

From the park to the library is ☐ blocks.

He traveled ☐ blocks.

Note taking

1n How much farther is the plant nursery from the burger shop than the smoothie shop is from the burger shop?

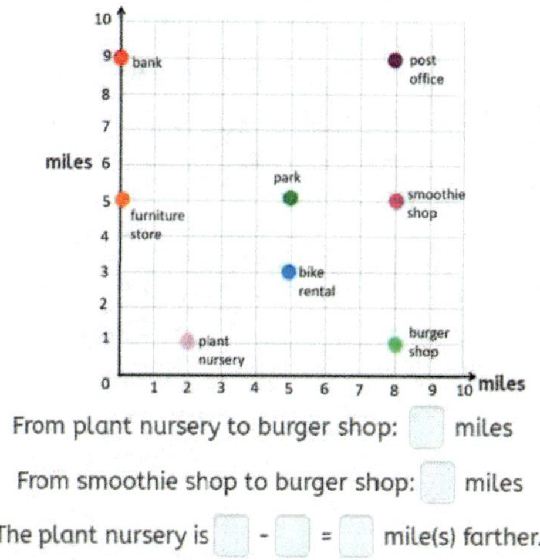

From plant nursery to burger shop: ☐ miles

From smoothie shop to burger shop: ☐ miles

The plant nursery is ☐ - ☐ = ☐ mile(s) farther.

Note taking

1o ...a walks from the bank to the furniture store and then to the park. How far is that?

From the bank to the furniture store: ☐ blocks

From the furniture store to the park: ☐ blocks

Tina walks ☐ blocks.

Note taking

1p ... gym will be built to form a rectangle with the other locations on the coordinate plane.

What will be the coordinates of the gym?

Steps

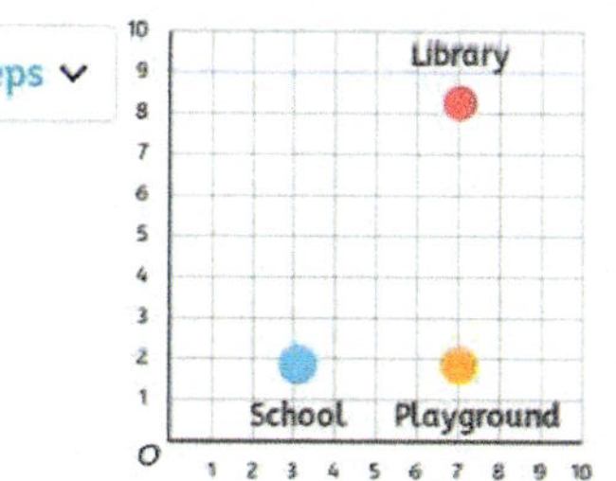

The gym will be ☐ units from the school and ☐ units from the library to make a rectangle.

The coordinates of the gym will be (☐, ☐).

Note taking

1q

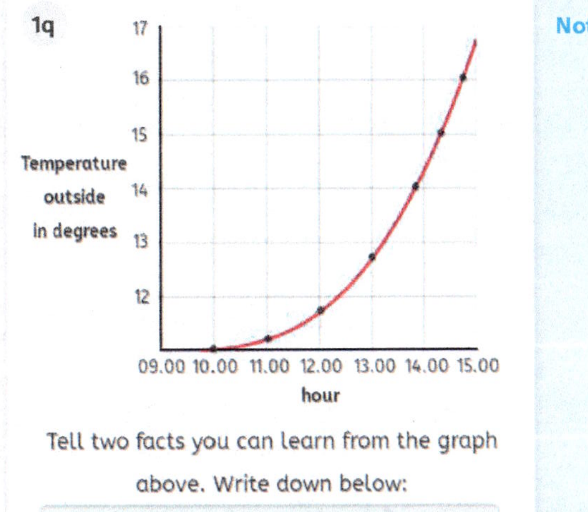

Tell two facts you can learn from the graph above. Write down below:

Note taking

2a Ling traveled from the school to the park and then to the library.

How many blocks did she travel?

Steps

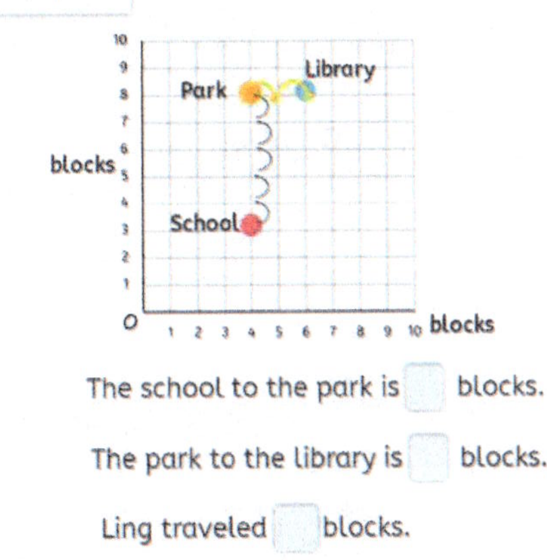

The school to the park is ☐ blocks.

The park to the library is ☐ blocks.

Ling traveled ☐ blocks.

Note taking

2b Ana left the restaurant and went to the gym. Then she went to the store.

How many blocks did she travel?

Steps

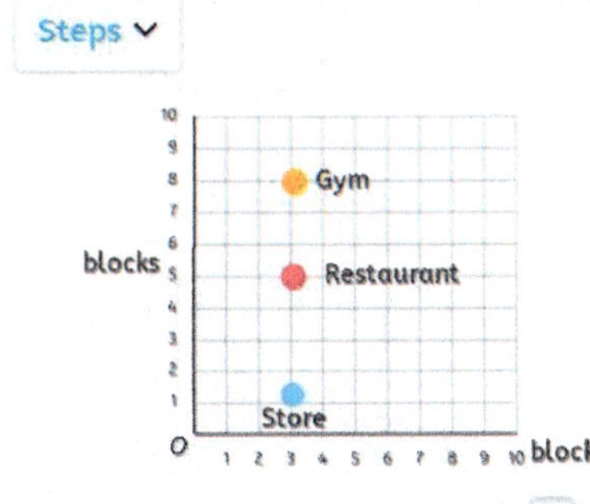

The restaurant to the gym is ☐ blocks.

The gym to the store is ☐ blocks.

Ana traveled ☐ blocks.

Note taking

2c Tina walks from Lonely Island to Smoky Buoy and then to Underwater Canyon. How far is that?

Steps

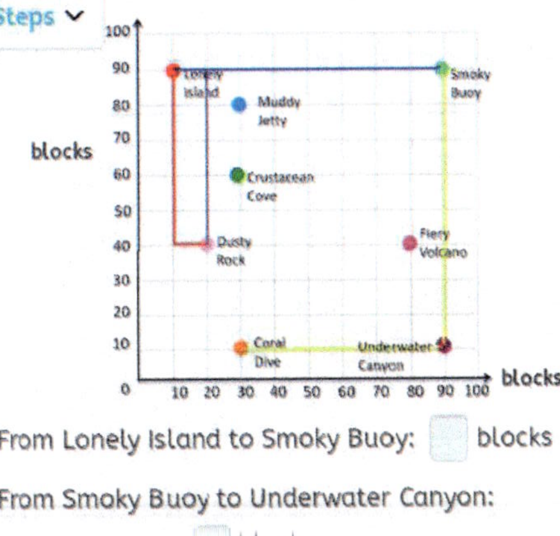

From Lonely Island to Smoky Buoy: ☐ blocks

From Smoky Buoy to Underwater Canyon: ☐ blocks

Tina walks ☐ blocks.

Note taking

2d How much farther is the park from the bank than the post office is from the bank?

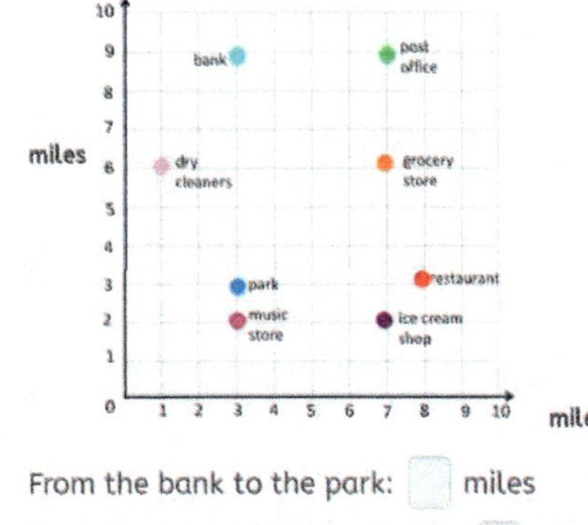

From the bank to the park: ☐ miles

From the bank to the post office: ☐ miles

The park is ☐ - ☐ = ☐ mile(s) farther.

Note taking

2e How far is it to go from the bank to the post office and then to the grocery store?

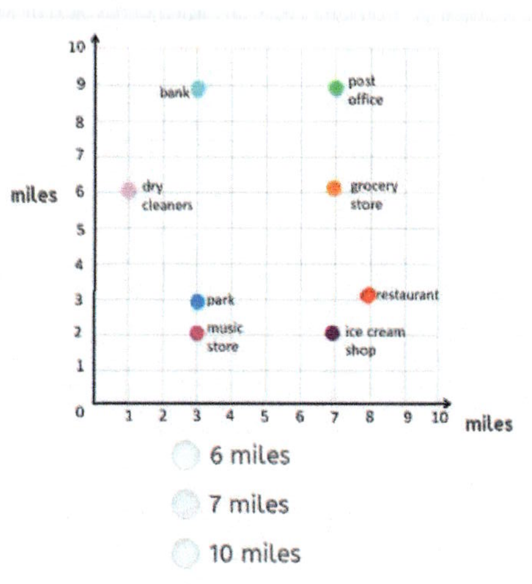

- 6 miles
- 7 miles
- 10 miles
- 14 miles

Note taking

2f David left the school located at (6, 2). He went to the park located at (2, 2). Then he went to the library located at (2, 8).

How many blocks did he travel?

Use the coordinate plane to help you.

From the school to the park is ☐ blocks.

From the park to the library is ☐ blocks.

He traveled ☐ blocks.

Note taking

2g Aiden left the school located at (4, 1). He went to the park located at (4, 4). Then he went to the library located at (0, 4).

How many blocks did he travel?

Use the coordinate plane to help you.

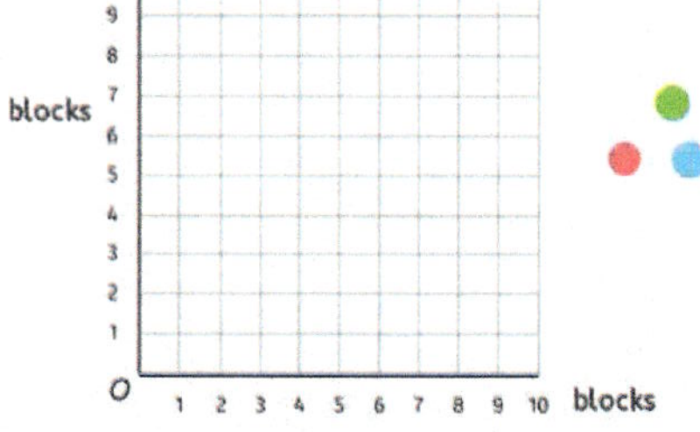

From the school to the park is ☐ blocks.

From the park to the library is ☐ blocks.

He traveled ☐ blocks.

Note taking

2h gym will be built to form a rectangle with the other locations on the coordinate plane.

What will be the coordinates of the gym?

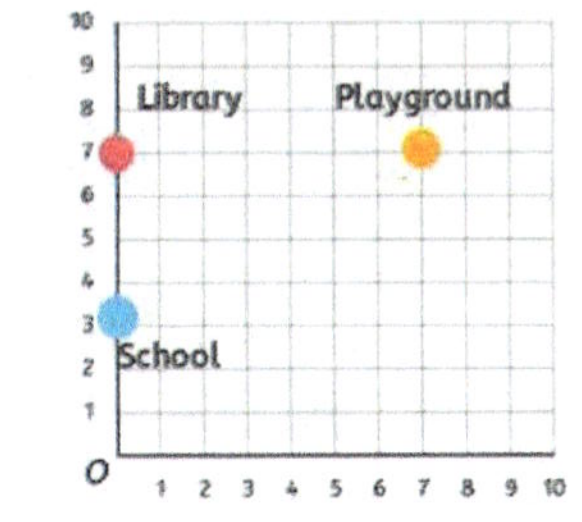

The gym will be ☐ units from the school and ☐ units from the playground to make a rectangle.

The coordinates of the gym will be (☐, ☐).

Note taking

2i ym and a store will be built so that all the locations form a square on the coordinate plane.

At which coordinates will the buildings be placed?

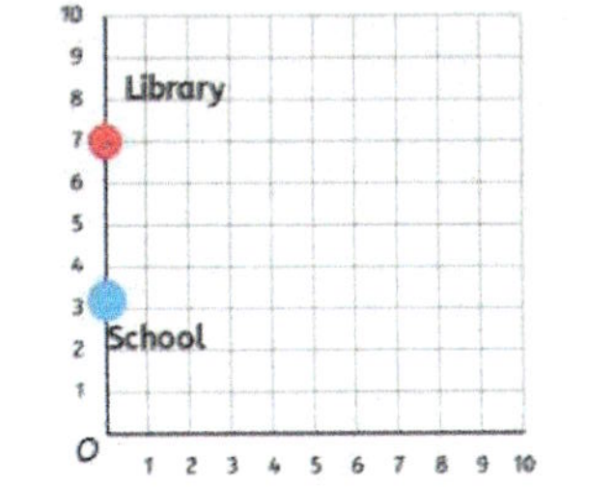

- (0, 3) and (0, 7)
- (4, 7) and (4, 3)
- (3, 4) and (7, 4)
- (3, 3) and (6, 6)

Note taking

2j ym and a store will be built so that all the locations form a square on the coordinate plane.

At which coordinates will the buildings be placed?

- (3, 1) and (3, 7)
- (1, 3) and (7, 3)
- (1, 9) and (7, 9)
- (9,1) and (9,7)

Note taking

2k T walks along horizontal and vertical lines by the shortest route to get from Coral Dive to Smoky Buoy. How far does she walk?

- 60 blocks
- 80 blocks
- 120 blocks
- 140 blocks

Note taking

2l

T... walks along horizontal and vertical lines by the shortest route to get from Muddy Jetty to Fiery Volcano. How far does she walk?

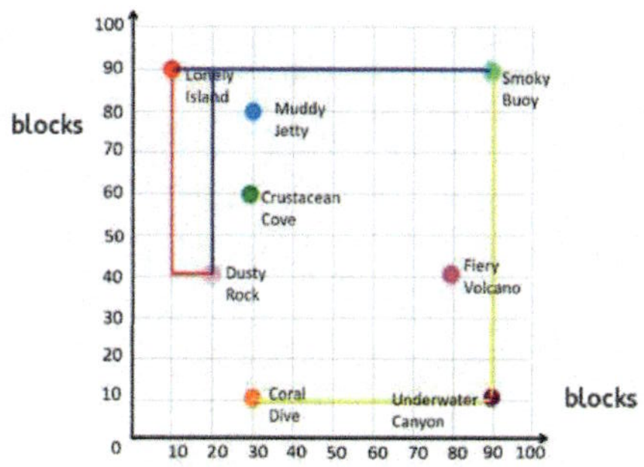

blocks
100
90
80
70
60
50
40
30
20
10
0
10 20 30 40 50 60 70 80 90 100
blocks

Lonely Island
Smoky Buoy
Muddy Jetty
Crustacean Cove
Dusty Rock
Fiery Volcano
Coral Dive
Underwater Canyon

- 40 blocks
- 50 blocks
- 90 blocks
- 100 blocks

Note taking

2m

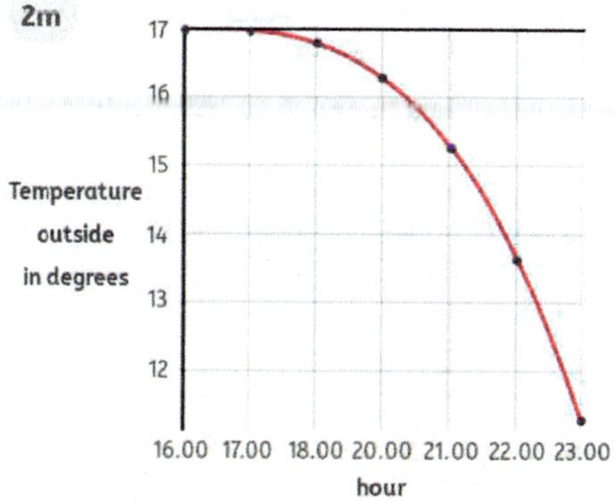

Temperature outside in degrees
17
16
15
14
13
12
16.00 17.00 18.00 20.00 21.00 22.00 23.00
hour

Tell two facts you can learn from the graph above. Write down below:

Note taking

1a Do you remember?

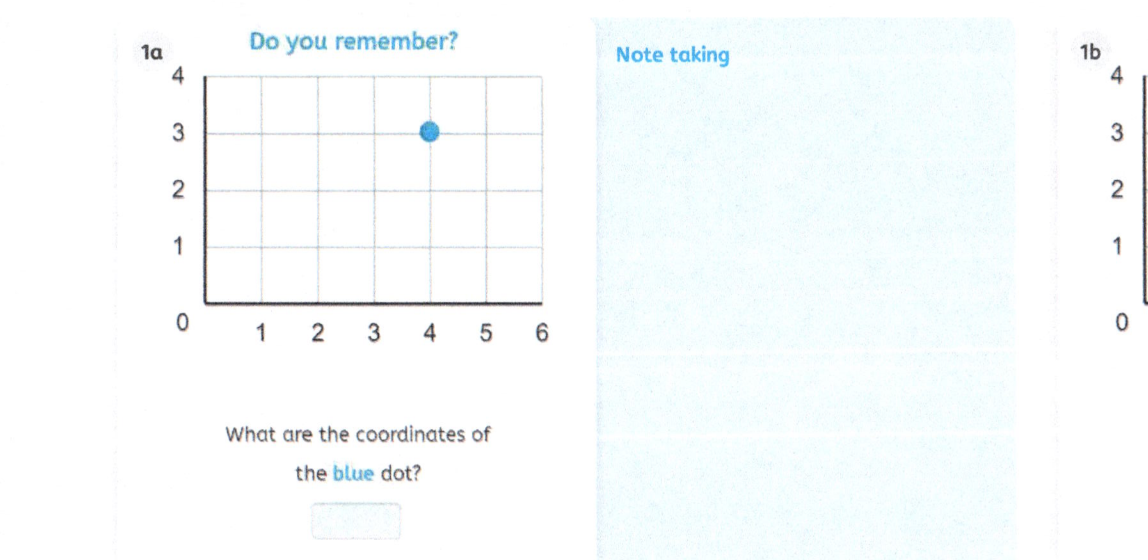

What are the coordinates of the blue dot?

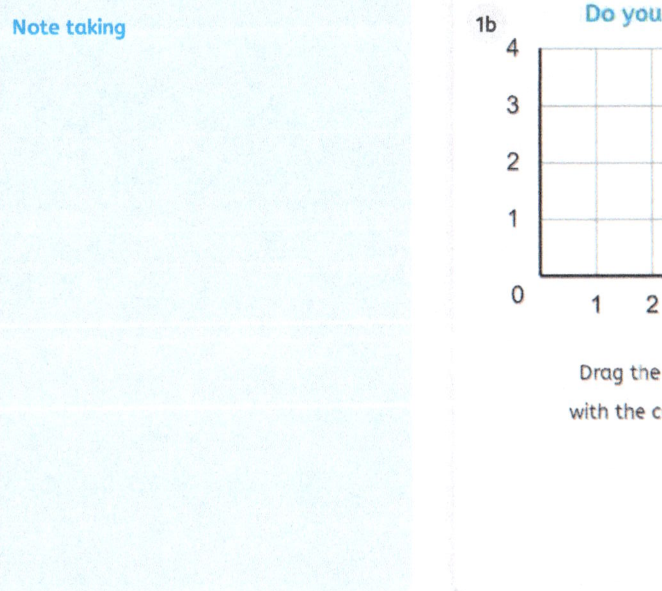

Note taking

1b Do you remember?

Drag the dot to the spot with the coordinates (5, 3).

Note taking

1l 3 for $2

Number of Bookmarks	Cost ($)
3	2
6	
	6

Ordered pairs:

(3,), (6,), (, 6)

Note taking

1m 3 for $2

Ordered Pairs: (3, 2), (6, 4), (9, 6)

Show the pattern on the coordinate plane.

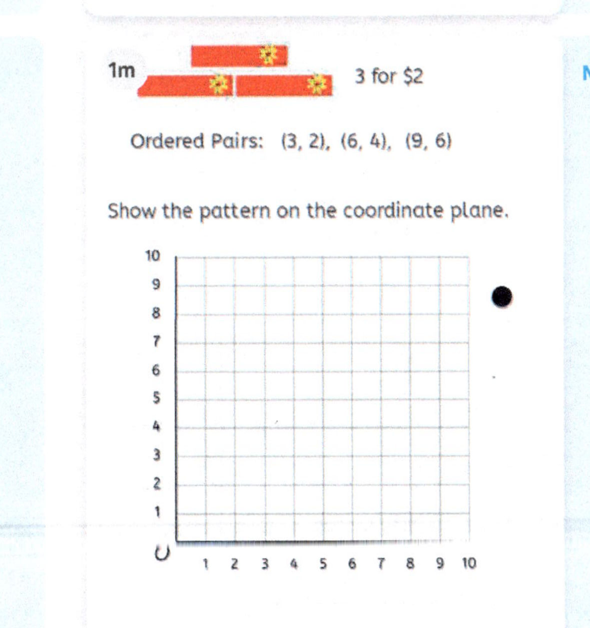

Note taking

1n

Pattern A: Add 2
Pattern B: Add 1

Complete the table.

Add 2	Add 1	Ordered Pair
2	1	(2, ☐
☐	2	(☐, ☐)
6	☐	(☐, ☐)

Note taking

1o

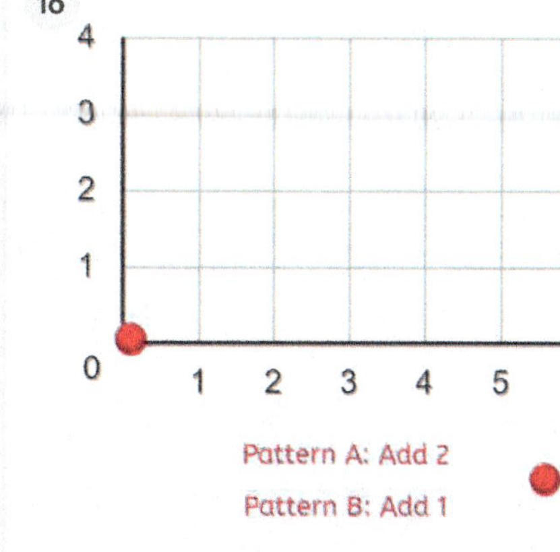

Pattern A: Add 2
Pattern B: Add 1

Plot the relationship between the patterns for (Pattern A, Pattern B).

Note taking

1p ·iden makes model cars and model trucks. It cost $5 to make each model car and $15 to make each model truck.

Number Made	Cost for Cars	Cost for Trucks	Ordered Pair
1	☐	15	(☐, ☐)
2	10	☐	(☐, ☐)
3	☐	☐	(☐, ☐)

The cost of making model trucks is ☐ times the cost of making model cars.

Note taking

2a 1 for $4

Number of Books	Cost ($)
1	4
2	☐
☐	12

Ordered pairs:

(1, ☐), (2, ☐), (☐, 12)

Note taking

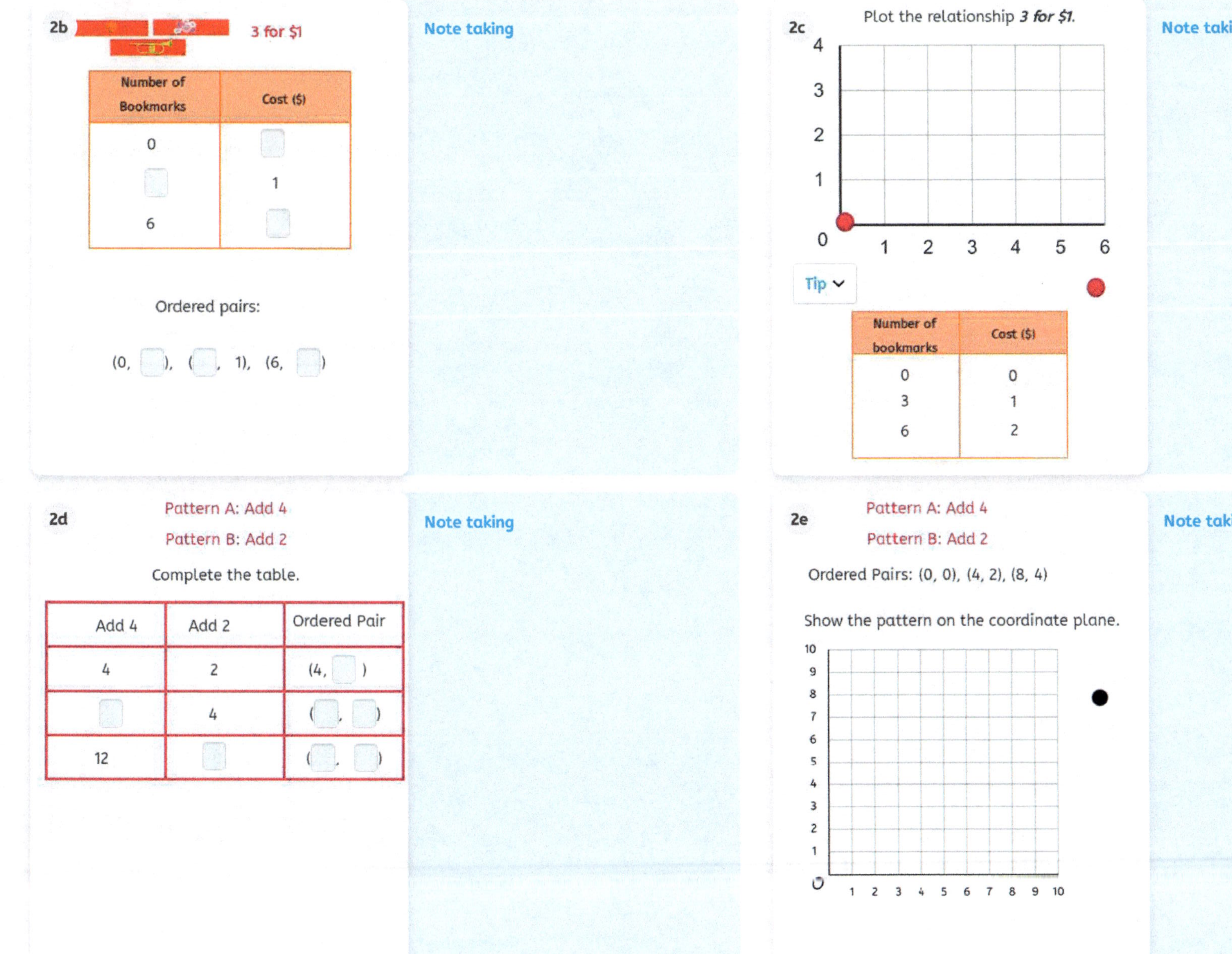

2b 3 for $1

Number of Bookmarks	Cost ($)
0	
	1
6	

Ordered pairs:

(0,), (, 1), (6,)

Note taking

2c Plot the relationship *3 for $1.*

Number of bookmarks	Cost ($)
0	0
3	1
6	2

Note taking

2d Pattern A: Add 4

Pattern B: Add 2

Complete the table.

Add 4	Add 2	Ordered Pair
4	2	(4,)
	4	(,)
12		(,)

Note taking

2e Pattern A: Add 4

Pattern B: Add 2

Ordered Pairs: (0, 0), (4, 2), (8, 4)

Show the pattern on the coordinate plane.

Note taking

2f

Pattern A: Add 3

Pattern B: Add 2

Plot the relationship between the patterns for (Pattern A, Pattern B).

Note taking

2g ...iden makes model cars and model trucks. It cost $4 to make each model car and $8 to make each model truck.

Number Made	Cost for Cars	Cost for Trucks	Ordered Pair
1		8	(,)
2	8		(,)
3			(,)

The cost of making model trucks is ___ times the cost of making model cars.

Note taking

2h Pencils are sold in packages of ten. Complete the ordered pairs that can be graphed to show the relationship between the number of packages and the number of pencils.

(0, ___), (___ , 10), (2, ___), (3, ___)

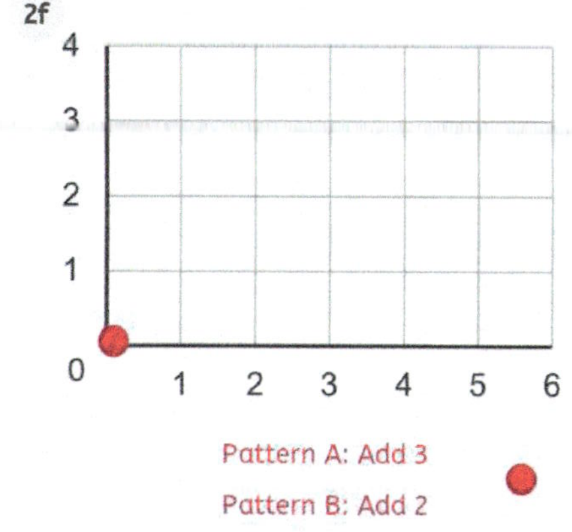

Note taking

2i ...nna earns $12 for every hour she works. Complete the ordered pairs that can be graphed to show the relationship between the number of hours Jenna works and the amount she earns.

(1, ___), (___ , 24), (3, ___), (4, ___)

Note taking

2j Mike earns $6 for every hour he works.
Jenna earns $12 for every hour she works.
Complete the ordered pairs that can be graphed to show the relationship between the amounts each person earns when they work the same number of hours.

(6, ☐), (☐, 24), (☐, 36), (24, ☐)

Jenna earns ☐ times the amount Mike earns.

Note taking

Made in the USA
Columbia, SC
30 August 2024